"十四五"职业教育国家规划教材

名校名师精品系列教材

Web Scraping with Python

# Python
## 爬虫项目教程
### 微课版

黄锐军 ◎ 主编

人民邮电出版社

北京

### 图书在版编目（CIP）数据

Python爬虫项目教程：微课版 / 黄锐军主编. -- 北京：人民邮电出版社，2021.11
名校名师精品系列教材
ISBN 978-7-115-56999-8

Ⅰ. ①P… Ⅱ. ①黄… Ⅲ. ①软件工具－程序设计－教材 Ⅳ. ①TP311.561

中国版本图书馆CIP数据核字(2021)第145956号

### 内 容 提 要

本书以 Python 语言为基础，介绍了爬虫的基础知识。本书包括 6 个实战项目，分别为爬取外汇网站数据、爬取名言网站数据、爬取电影网站数据、爬取图书网站数据、爬取商城网站数据、爬取景区网站数据。本书通过这些项目讲解了 Python 的 Web 访问技术、BeautifulSoup 的数据分析与提取技术、深度优先与广度优先顺序爬取技术、多线程网页爬取技术、scrapy 分布式爬取框架技术、selenium 爬虫程序技术、AWS 中的 DynamoDB 数据库的 NoSQL 存储技术等。本书每个项目都遵循由浅入深的学习规律，采取理论与实践相结合的方式来引导读者完成实战。

本书可作为计算机软件技术专业及其相关专业的程序设计课程教材。

◆ 主　编　黄锐军
　  责任编辑　初美呈
　  责任印制　王　郁　彭志环

◆ 人民邮电出版社出版发行　北京市丰台区成寿寺路 11 号
邮编　100164　电子邮件　315@ptpress.com.cn
网址　https://www.ptpress.com.cn
北京市鑫霸印务有限公司印刷

◆ 开本：787×1092　1/16
印张：16.75　　　　　　　　2021 年 11 月第 1 版
字数：438 千字　　　　　　 2024 年 12 月北京第 11 次印刷

定价：59.80 元

读者服务热线：(010)81055256　印装质量热线：(010)81055316
反盗版热线：(010)81055315
广告经营许可证：京东市监广登字 20170147 号

# 前言 PREFACE

爬虫程序指能自动从相关网页中搜索与提取所需数据的程序，提取与存储这些数据是进行数据分析的前提与关键。Python 语言因其简单、易读、可扩展的特性，在编写爬虫程序方面有特别的优势。尤其是业界有用 Python 编写的各种各样的爬虫程序框架供学习者参考，使得 Python 爬虫程序的编写更加简单、高效。

中国共产党二十大报告明确指出"教育、科技、人才是全面建设社会主义现代化国家的基础性、战略性支撑。""深化教育领域综合改革，加强教材建设和管理"。本书结合职业教育特色，以能力为导向，使用项目驱动的形式进行编写，让学生在做中学，在学中做。

本书包括 6 个项目。项目 1 基于爬取外汇网站数据，讲解 Python 的 Web 访问技术及正则表达式匹配字符串方法。项目 2 基于爬取名言网站数据，讲解 BeautifulSoup 的数据分析与提取技术。项目 3 基于爬取电影网站数据，讲解爬取多个网页数据的方法，重点讲解网页的深度优先与广度优先顺序爬取路径的构造方法与多线程网页爬取技术。项目 4 基于爬取图书网站数据，讲解目前功能强大的分布式爬取框架 scrapy 的程序技术。项目 5 基于爬取商城网站数据，讲解 selenium 爬虫程序技术。项目 6 基于爬取景区网站数据，讲解 AWS 中的 DynamoDB 数据库的 NoSQL 存储技术。每个项目都遵循由浅入深的学习规律，理论与实践相结合，引导读者完成实战。对于本书，建议分为 54 学时实施教学。

爬虫作为一种爬取网络数据的技术，同学们要正确使用网络爬虫技术。网络不是法外之地，爬虫不是牟利工具。同学们要遵守法律法规，合法有序地爬取数据，正确合理地使用爬取到的数据，不能危及网络安全，不能侵犯他人的知识产权。

由于编者知识水平有限，书中难免出现疏漏与不妥之处，欢迎读者批评指正。

编者
2022 年 11 月

# 目录

## 项目 1　爬取外汇网站数据 ……………… 1

- 1.1　项目任务 …………………………………… 1
- 1.2　搭建爬虫程序开发环境 ………………… 2
  - 1.2.1　理解爬虫程序 ……………………… 2
  - 1.2.2　搭建开发环境 ……………………… 2
- 1.3　使用 Flask 创建 Web 网站 …………… 3
  - 1.3.1　安装 Flask 框架 …………………… 3
  - 1.3.2　创建模拟外汇网站 ………………… 4
  - 1.3.3　编写客户端程序并获取网站的 HTML 代码 ………………………… 5
- 1.4　使用 GET 方法访问 Web 网站 ……… 7
  - 1.4.1　客户端使用 GET 方法发送数据 ………………………………… 7
  - 1.4.2　服务器端使用 GET 方法获取数据 ………………………………… 8
- 1.5　使用 POST 方法访问 Web 网站 …… 9
  - 1.5.1　客户端使用 POST 方法发送数据 ………………………………… 9
  - 1.5.2　服务器端使用 POST 方法获取数据 ………………………………… 10
  - 1.5.3　混合使用 GET 与 POST 方法 …… 11
- 1.6　使用正则表达式匹配数据 …………… 13
  - 1.6.1　使用正则表达式匹配字符串 …… 14
  - 1.6.2　使用正则表达式爬取数据 ……… 17
- 1.7　综合项目　爬取模拟外汇网站数据 ………………………………………… 18
  - 1.7.1　创建模拟外汇网站 ………………… 18
  - 1.7.2　解析网站的 HTML 代码 ………… 19
  - 1.7.3　设计存储数据库 ………………… 19
  - 1.7.4　编写爬虫程序 ……………………… 20
  - 1.7.5　执行爬虫程序 ……………………… 22
- 1.8　实战项目　爬取实际外汇网站数据 ………………………………………… 22
  - 1.8.1　解析网站的 HTML 代码 ………… 22
  - 1.8.2　爬取网站外汇汇率数据 ………… 24
  - 1.8.3　设计存储数据库 ………………… 25
  - 1.8.4　编写爬虫程序 ……………………… 26
  - 1.8.5　执行爬虫程序 ……………………… 28
- 项目总结 …………………………………………… 29
- 练习 1 ……………………………………………… 29

## 项目 2　爬取名言网站数据 ……………… 30

- 2.1　项目任务 ………………………………… 30
- 2.2　使用 BeautifulSoup 装载 HTML 文档 ………………………………………… 30
  - 2.2.1　创建模拟名言网站 ………………… 31
  - 2.2.2　安装 BeautifulSoup 程序包 ……… 32
  - 2.2.3　装载 HTML 文档 ………………… 32
- 2.3　使用 BeautifulSoup 查找 HTML 元素 ………………………………………… 34
  - 2.3.1　使用 find() 函数查找 …………… 34
  - 2.3.2　查找元素属性与文本 …………… 37
  - 2.3.3　使用 find_all() 函数查找 ……… 38
  - 2.3.4　使用高级查找 ……………………… 40
- 2.4　使用 BeautifulSoup 遍历文档元素 ………………………………………… 42
  - 2.4.1　获取元素节点的父节点 ………… 42
  - 2.4.2　获取元素节点的直接子节点 …… 43
  - 2.4.3　获取元素节点的所有子孙节点 ………………………………… 44
  - 2.4.4　获取元素节点的兄弟节点 ……… 45
- 2.5　BeautifulSoup 支持使用 CSS 语法进行查找 ………………………… 46

|  |  |  |  |
|---|---|---|---|
| | 2.5.1 使用 CSS 语法查找 ······ 47 | | 3.5.1 广度优先法 ······ 75 |
| | 2.5.2 使用属性的语法规则 ······ 48 | | 3.5.2 广度优先爬虫程序 ······ 76 |
| | 2.5.3 使用 select()查找子孙节点 ··· 49 | 3.6 | 爬取翻页网站数据 ······ 77 |
| | 2.5.4 使用 select()查找直接子节点 ··· 49 | | 3.6.1 使用 Flask 模板参数 ······ 77 |
| | 2.5.5 使用 select()查找兄弟节点 ··· 49 | | 3.6.2 创建翻页电影网站 ······ 81 |
| | 2.5.6 使用 select_one()查找单一元素 ······ 50 | | 3.6.3 编写爬虫程序 ······ 84 |
| | | | 3.6.4 执行爬虫程序 ······ 85 |
| 2.6 | 综合项目 爬取模拟名言网站数据 ······ 51 | 3.7 | 爬取网站全部图像 ······ 86 |
| | 2.6.1 创建模拟名言网站 ······ 51 | | 3.7.1 创建模拟电影网站 ······ 86 |
| | 2.6.2 爬取名言数据 ······ 51 | | 3.7.2 使用单线程程序爬取图像 ··· 88 |
| | 2.6.3 设计存储数据库 ······ 52 | | 3.7.3 使用 Python 的多线程 ······ 90 |
| | 2.6.4 编写爬虫程序 ······ 52 | | 3.7.4 使用多线程程序爬取图像 ··· 93 |
| | 2.6.5 执行爬虫程序 ······ 54 | 3.8 | 综合项目 爬取模拟电影网站数据 ······ 95 |
| 2.7 | 实战项目 爬取实际名言网站数据 ······ 55 | | 3.8.1 创建模拟电影网站 ······ 95 |
| | 2.7.1 解析网站的 HTML 代码 ······ 55 | | 3.8.2 设计存储数据库 ······ 98 |
| | 2.7.2 爬取全部页面的数据 ······ 56 | | 3.8.3 编写爬虫程序 ······ 99 |
| | 2.7.3 编写爬虫程序 ······ 57 | | 3.8.4 执行爬虫程序 ······ 102 |
| | 2.7.4 执行爬虫程序 ······ 59 | 3.9 | 实战项目 爬取实际电影网站数据 ······ 103 |
| 项目总结 ······ 60 | | | 3.9.1 解析电影网站的 HTML 代码 ······ 103 |
| 练习 2 ······ 60 | | | 3.9.2 爬取电影网站数据 ······ 105 |
| **项目 3 爬取电影网站数据** ······ 63 | | | 3.9.3 编写爬虫程序 ······ 107 |
| 3.1 | 项目任务 ······ 63 | | 3.9.4 执行爬虫程序 ······ 111 |
| 3.2 | 简单爬取网站数据 ······ 64 | 项目总结 ······ 112 | |
| | 3.2.1 创建模拟电影网站 ······ 65 | 练习 3 ······ 112 | |
| | 3.2.2 爬取网站数据 ······ 66 | **项目 4 爬取图书网站数据** ······ 113 | |
| | 3.2.3 编写爬虫程序 ······ 68 | 4.1 | 项目任务 ······ 113 |
| | 3.2.4 执行爬虫程序 ······ 69 | 4.2 | 使用 scrapy 创建爬虫程序 ······ 115 |
| 3.3 | 递归爬取网站数据 ······ 69 | | 4.2.1 创建网站服务器程序 ······ 115 |
| | 3.3.1 创建模拟电影网站 ······ 69 | | 4.2.2 安装 scrapy 框架 ······ 115 |
| | 3.3.2 解析电影网站结构 ······ 72 | | 4.2.3 scrapy 项目的创建 ······ 115 |
| | 3.3.3 递归爬取电影网站数据 ······ 72 | | 4.2.4 入口函数与入口地址 ······ 118 |
| 3.4 | 深度优先爬取网站数据 ······ 73 | | 4.2.5 Python 的 yield 语句 ······ 118 |
| | 3.4.1 深度优先法 ······ 73 | 4.3 | 使用 BeautifulSoup 爬取数据 ······ 119 |
| | 3.4.2 深度优先爬虫程序 ······ 74 | | 4.3.1 创建模拟图书网站 ······ 119 |
| 3.5 | 广度优先爬取网站数据 ······ 75 | | |

| | | | | |
|---|---|---|---|---|
| 4.3.2 | 解析网站的 HTML 代码 | 120 | 4.9.2 | 爬取网站图书数据 164 |
| 4.3.3 | 爬取图书图像 | 121 | 4.9.3 | 实现自动翻页 165 |
| 4.3.4 | 编写爬虫程序 | 122 | 4.9.4 | 编写爬虫程序 167 |
| 4.3.5 | 执行爬虫程序 | 123 | 4.9.5 | 执行爬虫程序并查看爬取结果 170 |

4.4 使用 XPath 查找元素 123

项目总结 172

4.4.1 scrapy 的 XPath 简介 124

练习 4 172

4.4.2 使用 XPath 查找 HTML 元素 125

## 项目 5 爬取商城网站数据 174

4.4.3 使用 XPath 与 BeautifulSoup 134

5.1 项目任务 174

4.5 爬取关联网页数据 135

5.2 使用 selenium 编写爬虫程序 176

4.5.1 创建模拟图书网站 135

5.2.1 JavaScript 程序控制网页 176

4.5.2 程序爬取网页的顺序 137

5.2.2 普通爬虫程序的问题 177

4.5.3 理解 scrapy 分布式 139

5.2.3 安装 selenium 与 Chrome 驱动程序 178

4.6 使用 XPath 爬取数据 140

5.2.4 编写 selenium 爬虫程序 178

4.6.1 创建模拟图书网站 140

5.3 使用 selenium 查找 HTML 元素 180

4.6.2 解析网站的 HTML 代码 142

5.3.1 创建模拟商城网站 180

4.6.3 爬取图书图像 143

5.3.2 使用 XPath 查找元素 182

4.6.4 设计数据库存储 144

5.3.3 查找元素的文本与属性 182

4.6.5 编写爬虫程序 144

5.3.4 使用 id 值查找元素 184

4.6.6 执行爬虫程序 146

5.3.5 使用 name 属性值查找元素 184

4.7 使用管道存储数据 147

5.3.6 使用 CSS 查找元素 184

4.7.1 创建模拟图书网站 147

5.3.7 使用 tagName 查找元素 185

4.7.2 编写数据字段类 149

5.3.8 使用文本查找超链接 186

4.7.3 编写爬虫程序类 150

5.3.9 使用 class 值查找元素 186

4.7.4 编写数据管道类 151

5.4 使用 selenium 实现用户登录 187

4.7.5 设置 scrapy 的配置文件 153

5.4.1 创建用户登录网站 187

4.7.6 执行爬虫程序 153

5.4.2 使用元素动作 188

4.8 综合项目 爬取模拟图书网站数据 154

5.4.3 编写爬虫程序 189

4.8.1 创建模拟图书网站 154

5.4.4 执行 JavaScript 程序 191

4.8.2 编写数据字段类 157

5.5 使用 selenium 爬取 Ajax 网页数据 192

4.8.3 编写数据管道类 157

5.5.1 创建 Ajax 网站 192

4.8.4 编写爬虫程序类 158

5.5.2 理解 selenium 爬虫程序 194

4.8.5 设置 scrapy 的配置文件 160

5.5.3 编写爬虫程序 197

4.8.6 执行爬虫程序 160

5.5.4 执行爬虫程序 198

4.9 实战项目 爬取实际图书网站数据 161

4.9.1 解析网站的 HTML 代码 161

5.6 使用 selenium 等待 HTML 元素 198
    5.6.1 创建延迟模拟网站 199
    5.6.2 编写爬虫程序 200
    5.6.3 selenium 强制等待 200
    5.6.4 selenium 隐式等待 201
    5.6.5 selenium 循环等待与显式等待 202
    5.6.6 selenium 显式等待形式 204
5.7 综合项目 爬取模拟商城网站数据 205
    5.7.1 创建模拟商城网站 205
    5.7.2 爬取网站数据并实现网页翻页 209
    5.7.3 设计数据存储与图像存储 210
    5.7.4 编写爬虫程序 211
    5.7.5 执行爬虫程序 214
5.8 实战项目 爬取实际商城网站数据 215
    5.8.1 解析网站的 HTML 代码 215
    5.8.2 爬取网站数据 218
    5.8.3 实现网页翻页 220
    5.8.4 编写爬虫程序 222
    5.8.5 执行爬虫程序 226
项目总结 228
练习 5 228

项目 6 爬取景区网站数据 230
6.1 项目任务 230
6.2 使用 DynamoDB 存储模拟景区网站数据 231
    6.2.1 创建模拟景区网站 231
    6.2.2 爬取网站数据 233
    6.2.3 编写爬虫程序 234
    6.2.4 执行爬虫程序 235
    6.2.5 DynamoDB 简介 235
6.3 登录 AWS 数据库 236
    6.3.1 登录 AWS 236
    6.3.2 创建数据库表 238
6.4 DynamoDB 数据库操作 240
    6.4.1 存储数据 240
    6.4.2 读取数据 241
    6.4.3 修改数据 242
    6.4.4 删除数据 243
    6.4.5 扫描数据 243
    6.4.6 删除数据库表 244
6.5 综合项目 爬取模拟景区网站数据 245
    6.5.1 创建模拟景区网站 245
    6.5.2 编写爬虫程序 245
    6.5.3 执行爬虫程序 248
6.6 实战项目 爬取实际景区网站数据 249
    6.6.1 解析网站的 HTML 代码 250
    6.6.2 爬取网站景区数据 252
    6.6.3 爬取全部页面的数据 254
    6.6.4 设计存储数据库 255
    6.6.5 编写爬虫程序 255
    6.6.6 执行爬虫程序 259
项目总结 260
练习 6 260

# 项目 1 爬取外汇网站数据

在很多网站中都有外汇汇率的数据,怎样得到这些数据呢?设计一个爬虫程序可以爬取这些数据。首先从网站中得到它的超文本标记语言(Hyper Text Markup Language,HTML)代码,然后解析相关 HTML 代码,最后提取所需的数据,这就是爬虫程序要完成的工作。爬虫程序的关键是解析与提取 HTML 代码中的数据,而正则表达式匹配是解析与提取 HTML 代码中数据的一种较为简单的方法。本项目通过爬取外汇网站数据的过程向读者介绍正则表达式的原理与应用。

我们在国内消费只需要人民币就行了,但是如果要进行进出口贸易,就需要换外汇了,换外汇就要知道人民币汇率。人民币汇率是人民币对某种外汇的比率或者兑换价,汇率变动对我国进出口贸易有着直接的调节作用。目前我国实行以市场供求为基础、参考一篮子货币进行调节、有管理的浮动汇率制度。通过对人民币汇率案例的学习,我们可以认识到中国货币政策的前瞻性,并理解中国货币政策的稳定性。

拓展阅读

人民币汇率变化对我国对我国进出口贸易的影响

## 1.1 项目任务

通过查询银行的网站可以看到各种外汇当前的汇率情况,如图 1-1-1 所示。我们的目标是设计爬虫程序,通过正则表达式匹配的方法爬取这些数据,并将其存储到数据库。

1-1-A 知识讲解    1-1-B 操作演练

在爬取这些数据之前,我们先练习爬取一个模拟网站的数据。通过 Flask 建立一个模拟外汇网站,如图 1-1-2 所示。该网站只有一个网页,其中展示了各种外汇的汇率数据。我们先学习编写一个爬虫程序,通过正则表达式匹配的方法爬取这个模拟外汇网站的汇率数据。

图 1-1-1 某时刻汇率情况      图 1-1-2 模拟外汇网站

## 1.2 搭建爬虫程序开发环境

1-2-A
知识讲解

1-2-B
操作演练

**任务目标**

认识爬虫程序，搭建爬虫程序开发环境。

图 1-2-1 所示是搭建好的 PyCharm 开发环境的程序界面。

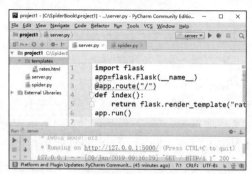

图 1-2-1　PyCharm 开发环境的程序界面

### 1.2.1　理解爬虫程序

爬虫程序是一组客户端程序，它的功能是访问 Web 服务器，从服务器中获取 HTML 代码，从中提取所需的数据，把数据整理后存储在模拟的数据库中。

例如，要想知道某天外汇的汇率数据，就要访问具有外汇汇率数据的网站，查看网站的 HTML 代码，代码中包含外汇汇率的数据，如图 1-2-2 所示。爬虫程序的作用就是获取 HTML 代码，从中解析出外汇汇率的数据，并将这些数据存储在数据库中。

图 1-2-2　网站的 HTML 代码

### 1.2.2　搭建开发环境

Python 是一种面向对象的解释型计算机程序设计语言，其具有以下特点：

- 开源、免费、功能强大；
- 语法简洁、清晰，强制用空白符表示语句缩进；
- 具有丰富和强大的库，能实现各种功能；
- 易读、易维护，用途广泛；
- 是解释型语言，变量类型可变，类似 JavaScript。

Python 自带一个集成开发环境（Integrated Development Environment，IDE），但是功能有限，此外还有很多第三方的 IDE。下面介绍几个主流的开发环境。

#### 1. Python 自带的开发环境

Python 的开发环境十分简单，读者可以到官网直接下载 Python 的程序包，几分钟就可以完成安装。安装完成后启动 Python，即可看到 Python 的命令行窗口。这个环境是命令行环境，只能运行一些简单的测试语句，不能用来编写程序。另外也可以启动 Python 自带的

IDE，但是这个 IDE 的功能十分有限，不适合开发 Python 工程项目。

### 2. PyCharm

一个比较流行的开发环境是 PyCharm，它是专门为 Python 开发的 IDE，具有很多功能，如调试、语法高亮、Project 管理、代码跳转、智能提示、自动完成、单元测试、版本控制等。读者可以到 PyCharm 的官网下载免费的 Community（社区）版本的 PyCharm，这个版本虽然不及收费的 Professional（专业）版本的 PyCharm 功能强大，但对于一般应用足够了。

### 3. Visual Studio Code

Visual Studio Code（简称 VS Code）是 Microsoft 开发的一个免费的 IDE，读者可以到 Microsoft 的官网下载后安装。VS Code 广泛地采用插件技术，在 VS Code 中安装 Python 的插件后可以进行 Python 的开发。VS Code 速度快，功能强大，也是一个很流行的 IDE。

### 4. Anaconda

另外一个比较流行的开发环境是 Anaconda。Anaconda 比较庞大，但却是一个十分强大的 Python 开发环境。它自带 Python 的解释器，也就是说，安装 Anaconda 时会自动安装 Python。同时，Anaconda 带有一个功能强大的 IDE——Spyder。Anaconda 最大的优点之一是可以帮助用户找到 Python 的各种库，从而使得 Python 的开发十分方便与高效。读者可以到官网下载 Anaconda。

PyCharm 比较小巧且功能强大，因此本书使用此开发环境。

## 1.3 使用 Flask 创建 Web 网站

1-3-A 知识讲解

1-3-B 操作演练

**任务目标**

读者先自己搭建一个 Web 网站，针对这个网站编写与调试爬虫程序，从而循序渐进地掌握爬虫的各种技术。Flask 是一个非常简单的 Python Web 开发框架，使用它可以快速搭建一个 Web 网站。

### 1.3.1 安装 Flask 框架

Python 的 Web 程序开发工具很多，Flask 是一个非常容易上手的 Python Web 开发框架，用户不需要知道太多关于 Web 编程的概念，只需要具备基本的 Python 开发技能就可以开发出一个 Web 应用。

在 Windows 操作系统中安装 Flask 非常简单，根据文档的介绍，直接在 Windows 命令提示符窗口执行以下命令即可。

```
pip install flask
```

如果显示下列信息，则表示 Flask 安装成功。

```
Successfully installed flask ...
```

关于 Flask 的更多信息，读者可以查看 Flask 的官网和 Flask 中文文档网站。

## 1.3.2 创建模拟外汇网站

### 1. 创建网站模板

通过 PyCharm 创建一个项目文件夹 project1，并在此文件夹中存储一个名称为 rates.csv 的文件。该文件的第一行是外汇汇率的价格描述信息，之后的每一行都是一种外币对应的汇率价格，各个部分用逗号分隔，内容如下：

```
交易币,交易币单位,现汇卖出价,现钞卖出价,现汇买入价,现钞买入价
新西兰元,100,466.28,466.28,462.56,447.93
澳大利亚元,100,488.99,488.99,485.09,469.75
美元,100,691.38,691.38,688.42,682.76
欧元,100,788.66,788.66,782.38,757.63
加拿大元,100,509.95,509.95,505.89,489.89
英镑,100,877.15,877.15,870.17,842.65
日元,100,6.2312,6.2312,6.1816,5.9861
新加坡元,100,504.49,504.49,500.47,484.64
瑞士法郎,100,697.09,697.09,691.53,669.66
```

### 2. 创建网站服务器

在 project1 中编写服务器程序 server.py，内容如下：

```python
import flask
app=flask.Flask(__name__)
@app.route("/")
def index():
    f=open("rates.csv","r",encoding="utf-8")
    st="<table border='1'>"
    rows=f.readlines()
    for row in rows:
        s=row.split(",")
        if len(s)==6:
            st=st+"<tr>"
            for t in s:
                st=st+"<td>"+t+"</td>"
            st=st+"</tr>"
    st=st+"</table>"
    f.close()
    return st
app.run()
```

### 3. 理解网站工作原理

这个网站的关键程序是服务器程序，现在来介绍其中语句和程序的功能。

（1）语句：

```
import flask
```

这条语句用于引入 flask 程序包，在 Flask 正确安装后都能正常引入该程序包。

（2）语句：

```
app=flask.Flask(__name__)
```

这条语句用于初始化一个 Flask 对象，参数__name__表示程序的名称。

（3）程序：

```
@app.route("/")
def index():
    f=open("rates.csv","r",encoding="utf-8")
    st="<table border='1'>"
    rows=f.readlines()
    for row in rows:
        s=row.split(",")
        if len(s)==6:
            st=st+"<tr>"
            for t in s:
                st=st+"<td>"+t+"</td>"
            st=st+"</tr>"
    st=st+"</table>"
    f.close()
    return st
```

@app.route("/")是路由控制语句，表示在访问根地址（相对地址是"/"）时就执行 index() 函数。这个函数用于打开 rates.csv 文件，读取每一行数据，然后通过","分解每行为一个列表，把列表的各个数据组织成一张表格，并返回该表格的 HTML 代码。

（4）语句：

```
app.run()
```

这条语句用于启动程序，执行 app.run() 函数就启动了一个 Web 服务器，它的默认地址是"http://127.0.0.1:5000"，因此访问该地址就执行 index() 函数返回表格的 HTML 代码。

运行这个服务器程序，可以看到出现一个地址为"http://127.0.0.1:5000"的网站，用浏览器访问这个地址，结果如图 1-3-1 所示。

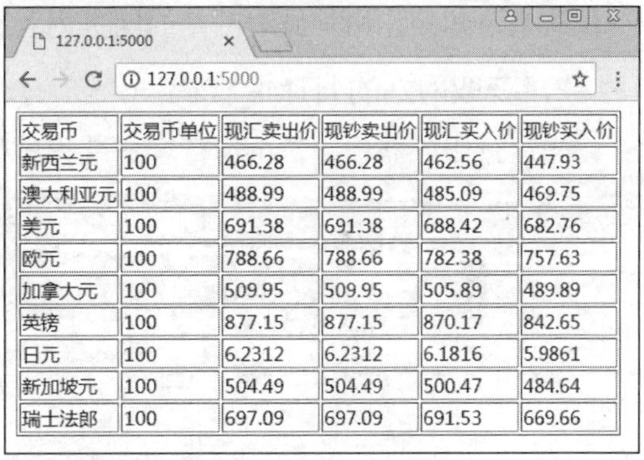

图 1-3-1 模拟外汇网站

## 1.3.3 编写客户端程序并获取网站的 HTML 代码

### 1. 编写客户端程序

这个网站除了可以使用浏览器访问外，也可以使用 urllib 程序包中的相关函数编写程序来访问。设计如下程序。

```
import urllib.request
url="http://127.0.0.1:5000"
html = urllib.request.urlopen(url)
html = html.read()
html = html.decode()
print(html)
```

运行程序后可以看到网站的 HTML 代码。下面来分析这个程序中部分语句的功能。

（1）语句：

```
import urllib.request
```

这条语句的作用是引入 urllib.request 程序包。这是 Python 自带的程序包，不需要安装。这个程序包的作用是访问网站。

（2）语句：

```
html = urllib.request.urlopen(url)
```

这条语句的作用是打开网站（这里打开的是自己的微型网站 http://127.0.0.1:5000），其中，urllib.request 是 urllib 库中的一个子程序包，urlopen()是用于打开网站的函数。

（3）语句：

```
html = html.read()
```

这个网站打开后要使用 read()函数读取网站的内容（如同读取文件的内容），注意读取的是二进制数据。

（4）语句：

```
html = html.decode()
```

这条语句的作用是把二进制数据转换为字符串，转换时默认的编码是 UTF-8。当然也可以通过语句指定转换编码，如 html=html.decode("utf-8")或 html=html.decode("gbk")，具体采用的编码需由网站的网页编码格式决定，如果编码不正确，则会出现中文乱码。

（5）语句：

```
print(html)
```

这条语句的作用是显示网站的网页内容，可见传递过来的就是网页数据。

### 2. 获取网站的 HTML 代码

先运行服务器程序，再运行客户端程序，可获取如下的网站 HTML 代码。

```
    <table border='1'><tr><td>交易币</td><td>交易币单位</td><td>现汇卖出价
</td><td>现钞卖出价</td><td>现汇买入价</td><td>现钞买入价
    </td></tr><tr><td>新西兰元</td><td>100</td><td>466.28</td><td>466.28
</td><td>462.56</td><td>447.93
    </td></tr><tr><td>澳大利亚元</td><td>100</td><td>488.99</td><td>488.99
</td><td>485.09</td><td>469.75
    </td></tr><tr><td>美元</td><td>100</td><td>691.38</td><td>691.38
</td><td>688.42</td><td>682.76
    </td></tr><tr><td>欧元</td><td>100</td><td>788.66</td><td>788.66
</td><td>782.38</td><td>757.63
    </td></tr><tr><td>加拿大元</td><td>100</td><td>509.95</td><td>509.95
</td><td>505.89</td><td>489.89
    </td></tr><tr><td>英镑</td><td>100</td><td>877.15</td><td>877.15
</td><td>870.17</td><td>842.65
    </td></tr><tr><td>日元</td><td>100</td><td>6.2312</td><td>6.2312
</td><td>6.1816</td><td>5.9861
    </td></tr><tr><td>新加坡元</td><td>100</td><td>504.49</td><td>504.49
</td><td>500.47</td><td>484.64
    </td></tr><tr><td>瑞士法郎</td><td>100</td><td>697.09</td><td>697.09
</td><td>691.53</td><td>669.66
    </td></tr></table>
```

# 项目 ❶  爬取外汇网站数据

获取网站的 HTML 代码是爬虫程序的第一个任务。实际上，HTML 代码中包含外汇的汇率数据，只是这些数据存放的位置比较凌乱，爬虫程序的功能就是在 HTML 代码中找到需要的外汇汇率数据，并把这些数据按一定格式存储。

## 1.4  使用 GET 方法访问 Web 网站

### 任务目标

前面创建的网站能返回所有外汇汇率,如果只关心某种外汇(如美元)汇率,就必须向网站服务器提交要查询的外汇名称。如图 1-4-1 所示,使用地址"http://127.0.0.1:5000/?currency=美元"访问网站即可查询到美元的汇率。在本节中,我们主要学习使用 GET 方法访问 Web 网站。

图 1-4-1  查询外汇汇率

### 1.4.1  客户端使用 GET 方法发送数据

#### 1. GET 方法简介

向网站发送数据常用的一种方法是 GET 方法,这种方法把要发送的数据放在地址的后面。例如,网站地址是"http://127.0.0.1:5000/",把参数"currency=美元"放在地址后面,成为"http://127.0.0.1:5000/?currency=美元",即向网站传递一个名称为 currency 的参数,该参数的值是"美元",地址与参数之间用"?"隔开。

使用 urllib 程序包编写客户端程序并向网站发送参数,采用的形式与直接用浏览器访问的形式很相似,只是如果参数值包含汉字,那么必须使用 urllib.parse.quote()函数对参数值进行编码,例如:

```
value=urllib.parse.quote("美元")
urllib.request.urlopen("http://127.0.0.1:5000?currency="+value)
```

如果不对汉字进行编码而直接使用汉字的值,就会出现错误。例如,下面是不正确的发送方式。

```
urllib.request.urlopen("http://127.0.0.1:5000?currency=美元")
```

#### 2. 客户端程序

编写的客户端程序如下:

```
import urllib.parse
import urllib.request
try:
```

```
        s=input("输入要查询的外汇:")
        value=urllib.parse.quote(s)
        html=urllib.request.urlopen("http://127.0.0.1:5000?currency="+value)
        html = html.read()
        html = html.decode()
        print(html)
    except Exception as err:
        print(err)
```

这个程序在访问"http://127.0.0.1:5000"时向网站传递了一个参数 currency，值是用户输入的外汇名称。

### 1.4.2 服务器端使用 GET 方法获取数据

#### 1. 服务器程序

客户端使用 GET 方法发送的数据存储在服务器的 flask.request.values 包中，服务器使用 GET 方法来获取参数。例如，服务器使用下面的语句就可以获取 GET 方法传递的 currency 的值。

```
currency=flask.request.values.get("currency")
```

但是如果客户端没有向服务器传递名称为 currency 的参数，那么这条语句会发生错误，因此在获取之前一般先判断一下。例如，使用下面的 if 语句进行判断。

```
if "currency" in flask.request.values:
    currency=flask.request.values.get("currency")
else:
    currency="美元"
```

如果 flask.request.values 中有 currency 参数，则使用 GET 方法获取，否则不能使用 GET 方法获取，默认设置 currency 的值为"美元"。

根据上面的分析，编写的服务器程序如下：

```
import flask
app=flask.Flask(__name__)
@app.route("/")
def index():
    if "currency" in flask.request.values:
        currency=flask.request.values.get("currency")
    else:
        currency="美元"
    f=open("rates.txt","r",encoding="utf-8")
    st="<table border='1'>"
    rows=f.readlines()
    i=0
    for row in rows:
        i=i+1
        s=row.split(",")
        if s[0]==currency or i==1:
            st=st+"<tr>"
            for t in s:
                st=st+"<td>"+t+"</td>"
            st=st+"</tr>"
```

```
                if i>1:
                    break
        st=st+"</table>"
        f.close()
        return st
    app.run()
```

程序中使用变量 i 来记录读到的行的编号，当 i=1 时代表读取的是标题行，因此将其加入 st 字符串中。如果某行的 s[0]为 currency，那么这一行就是要找的外汇，把它的数据加入 st 字符串中，最后把 st 组织成一张表格并返回。

**2. 运行程序**

运行客户端程序，输入要查询的外汇名称，如新西兰元，按"Enter"键后可以看到如下结果。

```
输入要查询的外币：新西兰元
<table border='1'><tr><td>交易币</td><td>交易币单位</td><td>现汇卖出价</td><td>现钞卖出价</td><td>现汇买入价</td><td>现钞买入价</td></tr>
    <tr><td>新西兰元</td><td>100</td><td>466.28</td><td>466.28</td><td>462.56</td><td>447.93
    </td></tr>
    </table>
```

实际上，程序的执行就是用"http://127.0.0.1:5000/?currency=新西兰元"网址访问网站的 HTML 代码。

## 1.5 使用 POST 方法访问 Web 网站

1-5-A　　1-5-B
知识讲解　操作演练

**任务目标**

向 Web 网站发送数据不仅有 GET 方法，还有 POST 方法。设计一个客户端程序，采用 POST 方法向网站发送要查询的外汇名称，然后网站返回该外汇的汇率，功能与使用 GET 方法查询一样。在本节中，我们主要学习使用 POST 方法访问 Web 网站。

### 1.5.1 客户端使用 POST 方法发送数据

前面介绍了使用 GET 方法查询外汇汇率，实际上，客户端向 Web 服务器传输参数还有一种常用的方法——POST 方法。POST 方法发送数据时与 GET 方法不同，它不把数据放在地址中，而是把数据包含在程序中，它采用的格式与 GET 方法的类似，但是要将字符串转换为二进制数据，例如：

```
value= "currency="+urllib.parse.quote("美元")
urllib.request.urlopen("http://127.0.0.1:5000",data=value.encode())
```

参数 currency 的值是"美元"，把它与 currency 组织成键值对后赋值给 value 变量，将 value 变量转换为二进制后传递给 urlopen()函数的 data 参数，这样 currency 参数就传递给了服务器。

根据这个规则编写的客户端程序如下：

```
import urllib.parse
```

```python
import urllib.request
try:
    s=input("输入要查询的外汇名称:")
    value=urllib.parse.quote(s)
    html=urllib.request.urlopen("http://127.0.0.1:5000",data=value.encode())
    html = html.read()
    html = html.decode()
    print(html)
except Exception as err:
    print(err)
```

**注意**：POST 方法的数据是放在 urlopen()函数的 data 参数中的，而不是放在地址后面的。

### 1.5.2 服务器端使用 POST 方法获取数据

服务器端使用 POST 方法获取数据的方式与使用 GET 方法获取数据的方式完全一样。获取 currency 参数的语句如下：

```
currency=flask.request.values.get("currency")
```

服务器程序唯一要修改的地方是路由语句。

```
@app.route("/",methods=["POST"])
```

这条语句表示该地址"/"可以响应 POST 请求，修改该语句是因为 Flask 的路由默认只响应 GET 请求，即下面两条语句是等效的。

```
@app.route("/")
@app.route("/",methods=["GET"])
```

在不指定 methods 时，Flask 只响应 GET 请求。如果要响应 POST 请求，就必须指定 methods=["POST"]。如果写成 methods=["GET","POST"]，就可以响应 GET 以及 POST 两种请求，例如：

```
@app.route("/",methods=["GET","POST"])
```

因此，编写的服务器程序如下：

```
import flask
app=flask.Flask(__name__)
@app.route("/",methods=["GET","POST"])
def index():
    if "currency" in flask.request.values:
        currency=flask.request.values.get("currency")
    else:
        currency="美元"
    f=open("rates.txt","r",encoding="utf-8")
    st="<table border='1'>"
    rows=f.readlines()
    i=0
    for row in rows:
        i=i+1
        s=row.split(",")
        if s[0]==currency or i==1:
            st=st+<tr>
            for t in s:
                st=st+<td>+t+</td>
            st=st+</tr>
```

```
                if i>1:
                    break
    st=st+"</table>"
    f.close()
    return st
app.run()
```

### 1.5.3 混合使用 GET 与 POST 方法

实际上，在应用中，客户端程序是可以同时使用 GET 方法与 POST 方法向服务器发送数据的。一般使用 GET 方法发送的数据，简单且数据量少，放在地址后面；而使用 POST 方法发送的数据，数据量大，放在程序中。

#### 1. 服务器程序

修改服务器程序，使其可以接收两个参数，一个是表示外汇名称的 currency 参数，另一个是表示返回格式的 format 参数，当 format 的值为"text"时，结果按文本格式返回，否则按表格格式返回。服务器程序如下：

```
import flask
app=flask.Flask(__name__)
def getTable(rows,currency):
    st = "<table border='1'>"
    i = 0
    for row in rows:
        i = i + 1
        s = row.split(",")
        if s[0] == currency or i == 1:
            st = st + <tr>
            for t in s:
                st = st + "<td>" + t + "</td>"
            st = st + </tr>
            if i > 1:
                break
    st = st + "</table>"
    return st

def getText(rows,currency):
    st =""
    i = 0
    for row in rows:
        i = i + 1
        s=row.split(",")
        if s[0] == currency or i == 1:
            st=st+row
            if i > 1:
                break
    return st

@app.route("/",methods=["GET","POST"])
def index():
    if "currency" in flask.request.values:
        currency=flask.request.values.get("currency")
```

```
        else:
            currency="美元"
    if "format" in flask.request.values:
        format=flask.request.values.get("format")
    else:
        format="table"
    f = open("rates.txt", "r", encoding="utf-8")
    rows = f.readlines()
    if format=="text":
        st=getText(rows,currency)
    else:
        st=getTable(rows,currency)
    f.close()
    return st
app.run()
```

这个程序接收 currency 参数与 format 参数,如果 format 的值为 "text",就调用 getText() 函数获取文本格式的结果,否则调用 getTable() 函数获取表格格式的结果。

### 2. 使用 GET 方法发送数据

使用 GET 方法发送的数据附加在地址后面,在地址后面接一个 "?",当有多个参数时,数据采用如下格式。

```
名称 1=值 1&名称 2=值 2&名称 3=值 3...
```

多个数据之间用 "&" 隔开,例如:

```
params= "currency="+urllib.parse.quote("美元")+"&format=text"
html=urllib.request.urlopen("http://127.0.0.1:5000?"+params)
```

其中,currency 与 format 两个参数之间用 "&" 隔开。由于 currency 的值是中文,因此使用 quote() 函数进行编码,而 format 的值是英文,不需要编码。

客户端程序如下:

```
import urllib.parse
import urllib.request
try:
    params= "currency="+urllib.parse.quote("美元")+"&format=text"
    html=urllib.request.urlopen("http://127.0.0.1:5000?"+params)
    html = html.read()
    html = html.decode()
    print(html)
except Exception as err:
    print(err)
```

先运行服务器程序,再运行客户端程序,返回文本格式的结果如下:

```
交易币,交易币单位,现汇卖出价,现钞卖出价,现汇买入价,现钞买入价
美元,100,691.48,691.48,688.42,682.76
```

如果把程序中的 format=text 改成 format=table,或去掉 format 参数,就返回表格格式的结果。

### 3. 使用 POST 方法发送数据

使用 POST 方法发送的数据采用与 GET 方法类似的格式,当有多个参数时,数据采用如下格式。

名称1=值1&名称2=值2&名称3=值3...

多个数据之间用"&"隔开,例如:

```
params= "currency="+urllib.parse.quote("美元")+"&format=text"
html=urllib.request.urlopen("http://127.0.0.1:5000",data=params.encode())
```

其中,currency 与 format 两个参数之间用"&"隔开。由于 currency 的值是中文,因此使用 quote()函数进行编码,而 format 的值是英文,不需要编码。

客户端程序如下:

```
import urllib.parse
import urllib.request
try:
    params= "currency="+urllib.parse.quote("美元")+"&format=text"
    html=urllib.request.urlopen("http://127.0.0.1:5000",data=params.encode())
    html = html.read()
    html = html.decode()
    print(html)
except Exception as err:
    print(err)
```

#### 4. 同时使用 GET 与 POST 方法发送数据

一般来说,客户端程序可以同时使用 GET 与 POST 方法发送数据,只是使用 GET 方法发送的数据放在地址中,而使用 POST 方法发送的数据放在 urlopen()函数的 data 参数中。例如,客户端程序可以编写如下:

```
import urllib.parse
import urllib.request
try:
    params=( "currency="+urllib.parse.quote("美元")).encode()
    html=urllib.request.urlopen("http://127.0.0.1:5000?format=text",data=params)
    html = html.read()
    html = html.decode()
    print(html)
except Exception as err:
    print(err)
```

该程序与上述客户端程序的效果是一样的,都是向服务器传递了 currency="美元"与 format="text"参数,只是 format 参数用 GET 方法传递,而 currency 参数用 POST 方法传递。

## 1.6 使用正则表达式匹配数据

**任务目标**

前面创建了一个模拟外汇网站,可以查询某种外汇的汇率。现在来编写一个爬虫程序,使它可以爬取指定外汇汇率的格式化数据。要得到这些数据,就必须从返回的 HTML 代码中解析出每个单独的数据,而一种比较简单的解析方法就是采用正则表达式。

1-6-A  1-6-B
知识讲解  操作演练

## 1.6.1 使用正则表达式匹配字符串

正则表达式是用来匹配与查找字符串的,从网上爬取数据或多或少都会用到正则表达式。Python 的正则表达式要引入 re 程序包,以 r 引导,例如:

```
import re
reg=r"\d+"
m=re.search(reg,"abc123cd")
print(m)
```

其中,r"\d+"表示匹配连续的多个数值,search()是 re 程序包中的函数,用于从 "abc123cd" 字符串中搜索连续的数值,搜索到 "123" 后,程序返回一个匹配对象,结果如下。

```
<_sre.SRE_Match object; span=(3, 6), match='123'>
```

从结果可以看出,程序在指定的字符串中找到了连续的数值,即 "123"。span(3,6)表示数值的开始位置是 3,结束位置是 6,这正好是 "123" 在 "abc123cd" 中的位置。

Python 中关于正则表达式的规则比较多,下面将介绍主要的内容,详细内容读者可以参考相关资料。

(1)字符 "\d" 匹配 0~9 的一个数值。

例如:

```
import re
reg=r"\d"
m=re.search(reg,"abc123cd")
print(m)
```

结果:

```
<_sre.SRE_Match object; span=(3, 4), match='1'>
```

匹配结果是第一个数值 "1"。

(2)字符 "+" 可以重复前面的一个匹配字符一次或者多次。

例如:

```
import re
reg=r"b\d+"
m=re.search(reg,"a12b123c")
print(m)
```

结果:

```
<_sre.SRE_Match object; span=(3, 7), match='b123'>
```

匹配结果是 "b123"。

**注意**:r"b\d+"表示第一个字符要匹配 "b",后面是连续的多个数值,因此匹配的是 "b123",不是 "a12"。

(3)字符 "*" 表示重复前面的一个匹配字符零次或者多次。

"*" 与 "+" 类似,但有区别,例如:

```
import re
reg=r"ab+"
m=re.search(reg,"acabc")
print(m)
reg=r"ab*"
m=re.search(reg,"acabc")
print(m)
```

结果:
```
<_sre.SRE_Match object; span=(2, 4), match='ab'>
<_sre.SRE_Match object; span=(0, 1), match='a'>
```
由此可见，r"ab+"匹配的是"ab"，而r"ab*"匹配的是"a"，因为r"ab*"表示"b"可以重复零次，而r"ab+"却要求"b"重复一次以上。

（4）字符"?"表示重复前面的一个匹配字符零次或一次。

例如:
```
import re
reg=r"ab?"
m=re.search(reg,"abbcabc")
print(m)
```
结果:
```
<_sre.SRE_Match object; span=(0, 2), match='ab'>
```
匹配结果是"ab"，其中，"b"重复一次。

（5）字符"."代表任意字符，但是没有特别声明时不代表字符"\n"。

例如:
```
import re
s="xaxby"
m=re.search(r"a.b",s)
print(m)
```
结果:
```
<_sre.SRE_Match object; span=(1, 4), match='axb'>
```
"."代表了字符"x"。

（6）"|"表示|符号左右两个部分任意匹配一个。

例如:
```
import re
s="xaababaaby"
m=re.search(r"ab|ba",s)
print(m)
```
匹配"ab"或者"ba"都可以，结果:
```
<_sre.SRE_Match object; span=(2, 4), match='ab'>
```
（7）特殊字符使用"\"引导，例如，"\r""\n""\t""\\"分别表示回车符、换行符、制表符与反斜线本身。

例如:
```
import re
reg=r"a\nb?"
m=re.search(reg,"ca\nbcabc")
print(m)
```
结果:
```
<_sre.SRE_Match object; span=(1, 4), match='a\nb'>
```
匹配结果是"a\nb"。

（8）字符"\b"表示单词结尾，各种空白字符或者字符串结尾都是单词结尾。

例如:
```
import re
```

```
reg=r"car\b"
m=re.search(reg,"The car is black")
print(m)
```

结果：

```
<_sre.SRE_Match object; span=(4, 7), match='car'>
```

匹配结果是"car"，因为"car"后面是一个空格。

（9）"[]"中的字符可以任选一个，如果字符是 ASCII 中连续的一组，那么可以使用"-"连接。例如，[0-9]表示 0～9 中的一个数字，[A-Z]表示 A～Z 中的一个大写字母，[0-9A-Z]表示 0～9 中的一个数字或 A～Z 中的一个大写字母。

例如：

```
import re
reg=r"x[0-9]y"
m=re.search(reg,"xyx2y")
print(m)
```

结果：

```
<_sre.SRE_Match object; span=(2, 5), match='x2y'>
```

匹配结果是"x2y"。

（10）"^"出现在"[]"中的第一个字符的位置，代表取反。例如，[^ab0-9]表示不是 a、b，也不是 0～9 的数字。

例如：

```
import re
reg=r"x[^ab0-9]y"
m=re.search(reg,"xayx2yxcy")
print(m)
```

结果：

```
<_sre.SRE_Match object; span=(6, 9), match='xcy'>
```

匹配结果是"xcy"。

（11）"\s"匹配任何空白字符，等价于"[\r\n\x20\t\f\v]"。

例如：

```
import re
s="1a ba\tbxy"
m=re.search(r"a\sb",s)
print(m)
```

结果：

```
<_sre.SRE_Match object; span=(1, 4), match='a b'>
```

匹配结果是"a b"。

（12）"\w"匹配下画线、英文、数字等字符，等价于"[a-zA-Z0-9_]"。

例如：

```
import re
reg=r"\w+"
m=re.search(reg,"Python is easy")
print(m)
```

结果：

```
<_sre.SRE_Match object; span=(0, 6), match='Python'>
```

匹配结果是"Python"。

（13）"^"匹配字符串的开头位置。

例如：

```
import re
reg=r"^ab"
m=re.search(reg,"cabcab")
print(m)
```

结果：

```
None
```

没有匹配到任何字符，因为"cabcab"中虽然有"ab"，但不以"ab"开头。

（14）"$"匹配字符串的结尾位置。

例如：

```
import re
reg=r"ab$"
m=re.search(reg,"abcab")
print(m)
```

结果：

```
<_sre.SRE_Match object; span=(3, 5), match='ab'>
```

**注意**：匹配的是最后一个"ab"，而不是第一个"ab"。

（15）"()"经常与"+""*""?"等一起使用，对"()"内的部分进行重复。

例如：

```
import re
reg=r"(ab)+"
m=re.search(reg,"ababcab")
print(m)
```

结果：

```
<_sre.SRE_Match object; span=(0, 4), match='abab'>
```

匹配结果是"abab"，因为"+"对"ab"进行了重复。

### 1.6.2 使用正则表达式爬取数据

如果有一组数据包含在下面的表格中：

```
<table><tr><td>美元</td><td>100</td><td>691.38</td><td>691.38</td><td>688.42</td><td>682.76
</td></tr></table>
```

怎样提取其中的数据呢？显然可以使用正则表达式。正则表达式库 re 程序包中的 search()函数使用正则表达式对要匹配的字符串进行匹配，如果匹配不成功就返回 None，如果匹配成功就返回一个匹配对象。匹配对象调用 start()函数得到匹配字符串的开始位置，匹配对象调用 end()函数得到匹配字符串的结束位置。search()函数虽然只返回第一次匹配的结果，但是只要连续使用 search()函数，就可以找到字符串中全部匹配的内容。

例如，解析字符串：

```
<tr><td>美元</td><td>100</td><td>691.38</td><td>691.38</td><td>688.42
</td><td>682.76</td></tr>
```

可以首先匹配<tr>与</tr>，找到它们的位置，再提取中间部分：

```
<td>美元</td><td>100</td><td>691.38</td><td>691.38</td><td>688.42
</td><td>682.76</td>
```

接着匹配第一个<td>与</td>，提取中间的部分"美元"，再把字符串缩减为：

```
<td>100</td><td>691.38</td><td>691.38</td><td>688.42</td><td>682.76</td>
```

之后匹配第二个<td>与</td>，提取其中的数据，以此类推，就可以得到所有数据。

根据上面的分析，编写一个实验程序。

```python
import re
s="<tr><td>美元</td><td>100</td><td>691.38</td><td>691.38</td><td>688.42</td><td>682.76</td></tr>
"
m=re.search(r"<tr>",s)
n=re.search(r"</tr>",s)
s=s[m.end():n.start()]
while s!="":
    m = re.search(r"<td>", s)
    n = re.search(r"</td>", s)
    t = s[m.end():n.start()]
    print(t)
    s=s[n.end():]
```

该程序先匹配<tr>与</tr>，通过 s=s[m.end():n.start()]语句将<tr>…</tr>的中间部分提取，再匹配<td>与</td>，<td>与</td>中包含的就是要提取的数据，通过 s=s[n.end():]语句缩减字符串，直到字符串缩减为空，从而得到所有数据。

执行该程序，结果如下：

```
美元
100
691.38
691.38
688.42
682.76
```

结果表明，该程序已经解析出所有数据。

## 1.7 综合项目 爬取模拟外汇网站数据

1-7-A 知识讲解
1-7-B 操作演练

**任务目标**

创建一个模拟外汇网站，编写爬虫程序以获取网站的 HTML 代码，使用正则表达式解析 HTML 代码，爬取外汇汇率数据并将其存储到数据库。

### 1.7.1 创建模拟外汇网站

使用 project1 文件夹下 rates.csv 文件中的数据，编写服务器程序 server.py。

```python
import flask
app=flask.Flask(__name__)
@app.route("/")
def index():
```

```
        f=open("rates.csv","r",encoding="utf-8")
        st="<table border='1'>"
        rows=f.readlines()
        for row in rows:
            s=row.split(",")
            if len(s)==6:
                st=st+"<tr>"
                for t in s:
                    st=st+"<td>"+t+"</td>"
                st=st+"</tr>"
        st=st+"</table>"
        f.close()
        return st
app.run()
```

运行这个服务器程序，可以看到出现一个地址为"http://127.0.0.1:5000"的网站，用浏览器访问这个地址，结果如图1-3-1所示。

### 1.7.2 解析网站的HTML代码

使用Chrome浏览器访问"http://127.0.0.1:5000"，找到外汇汇率的其中一行数据（如美元），单击鼠标右键，在弹出的快捷菜单中选择"检查"命令，就可以看到对应的HTML代码，如图1-7-1所示。

在HTML代码中，数据被包含在<table>…</table>中，第一行<tr>…</tr>表示表格标题，之后的每行<tr>…</tr>都包含一组外汇的汇率数据，如美元的数据行：

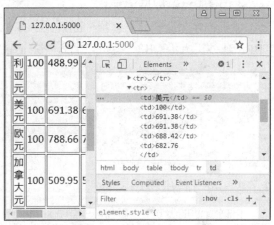

图 1-7-1 网站的 HTML 代码

```
<tr><td>美元</td><td>100</td><td>691.38</td><td>691.38</td><td>688.42</td><td>682.76
</td></tr>
```

可以使用正则表达式爬取这些数据。

### 1.7.3 设计存储数据库

#### 1. 设计数据库表

设计一个SQLite3数据库rates.db，它有一张rates表，其结构如表1-7-1所示。

表 1-7-1 rates 表结构

| 字段名称 | 类型 | 说明 |
| --- | --- | --- |
| Currency | varchar(256) | 外汇名称（关键字） |
| TSP | float | 现汇卖出价 |
| CSP | float | 现钞卖出价 |
| TBP | float | 现汇买入价 |
| CBP | float | 现钞买入价 |

### 2. 创建数据库表

设计一个 MySpider 爬虫类，创建一个 openDB(self)函数，连接数据库 rates.db 并创建 rates 表，代码如下：

```python
def openDB(self):
    self.con=sqlite3.connect("rates.db")
    self.cursor=self.con.cursor()
    try:
        self.cursor.execute("drop table rates")
    except:
        pass
    sql="create table rates (Currency varchar(256) primary key,TSP float,CSP float, TBP float, CBP float)"
    self.cursor.execute(sql)
```

**注意**：如果 rates 表已经存在就将其删除，然后创建一个空的 rates 表。

### 3. 增加一条数据记录

如果程序爬取到外汇名称 Currency、现汇卖出价 TSP、现钞卖出价 CSP、现汇买入价 TBP、现钞买入价 CBP，就可以插入一条记录。insertDB()函数可完成这个插入工作。

```python
def insertDB(self, Currency, TSP, CSP, TBP, CBP):
    try:
        sql = "insert into rates (Currency,TSP,CSP,TBP,CBP) values (?,?,?,?,?)"
        self.cursor.execute(sql, [Currency, TSP, CSP, TBP, CBP])
    except Exception as err:
        print(err)
```

## 1.7.4 编写爬虫程序

根据前面的分析，编写爬虫程序 spider.py。

```python
import urllib.request
import re
import sqlite3

class MySpider:
    def openDB(self):
        self.con=sqlite3.connect("rates.db")
        self.cursor=self.con.cursor()
        try:
            self.cursor.execute("drop table rates")
        except:
            pass
        sql="create table rates (Currency varchar(256) primary key,TSP float,CSP float, TBP float, CBP float)"
        self.cursor.execute(sql)

    def closeDB(self):
        self.con.commit()
        self.con.close()

    def insertDB(self,Currency,TSP,CSP,TBP,CBP):
```

```python
            try:
                sql="insert into rates (Currency,TSP,CSP,TBP,CBP) values (?,?,?,?,?)"
                self.cursor.execute(sql,[Currency,TSP,CSP,TBP,CBP])
            except Exception as err:
                print(err)

    def show(self):
        self.cursor.execute("select Currency,TSP,CSP,TBP,CBP from rates")
        rows=self.cursor.fetchall()
        print("%-18s%-12s%-12s%-12s%-12s" %("Currency","TSP","CSP","TBP","CBP"))
        for row in rows:
            print("%-18s%-12.2f%-12.2f%-12.2f%-12.2f" % (row[0],row[1],row[2],row[3],row[4]))

    def spider(self,url):
        try:
            resp=urllib.request.urlopen(url)
            data=resp.read()
            html=data.decode()
            p=re.search(r"<tr>",html)
            q=re.search(r"</tr>",html)
            i=0
            while p and q:
                a=p.end()
                b=q.start()
                tds=html[a:b]
                m=re.search(r"<td>",tds)
                n=re.search(r"</td>", tds)
                row=[]
                while m and n:
                    u=m.end()
                    v=n.start()
                    row.append(tds[u:v].strip("\n"))
                    tds=tds[n.end():]
                    m=re.search(r"<td>",tds)
                    n=re.search(r"</td>", tds)
                i=i+1
                if i>=2 and len(row)==6:
                    Currency=row[0]
                    TSP=float(row[2])
                    CSP=float(row[3])
                    TBP=float(row[4])
                    CBP=float(row[5])
                    self.insertDB(Currency,TSP,CSP,TBP,CBP)
                html=html[q.end():]
                p=re.search(r"<tr>",html)
                q=re.search(r"</tr>",html)
        except Exception as err:
            print(err)

    def process(self):
        self.openDB()
```

```
            self.spider("http://127.0.0.1:5000")
            self.show()
            self.closeDB()
#主程序
spider=MySpider()
spider.process()
```

其中，spider()函数中使用 p = re.search(r"<tr>", html)以及 q = re.search(r"</tr>", html)语句匹配<tr>与</tr>。如果匹配到<tr>与</tr>，就使用 tds = html[p.end():q.start()]语句获取其中的字符串，并使用 html=html[q.end():]语句获取待分析的</tr>后面的字符串；如果匹配不到<tr>与</tr>，就说明已经没有数据了，退出外层循环。

程序在使用 tds = html[p.end():q.start()]语句获取 HTML 字符串中间的字符串后，进一步使用 m = re.search(r"<td>", tds)语句与 n = re.search(r"</td>", tds)语句匹配<td>与</td>，每次匹配到一个元素就使用 tds[m.end():n.start()].strip("\n")语句提取该元素的值。如果匹配不到<td>与</td>，则表示一行的数据已经爬取完毕，退出内层循环。

程序使用变量 i 记录行号，第 1 行是标题，从第 2 行开始就是数据，获取外汇名称 Currency、现汇卖出价 TSP、现钞卖出价 CSP、现汇买入价 TBP、现钞买入价 CBP，调用 insertDB()函数把数据插入数据库。

### 1.7.5　执行爬虫程序

先执行服务器程序，再执行爬虫程序，将爬取结果存储到数据库，通过 show()函数显示的结果如图 1-7-2 所示。

| Currency | TSP | CSP | TBP | CBP |
| --- | --- | --- | --- | --- |
| 新西兰元 | 466.28 | 466.28 | 462.56 | 447.93 |
| 澳大利亚元 | 488.99 | 488.99 | 485.09 | 469.75 |
| 美元 | 691.38 | 691.38 | 688.42 | 682.76 |
| 欧元 | 788.66 | 788.66 | 782.38 | 757.63 |
| 加拿大元 | 509.95 | 509.95 | 505.89 | 489.89 |
| 英镑 | 877.15 | 877.15 | 870.17 | 842.65 |
| 日元 | 6.23 | 6.23 | 6.18 | 5.99 |
| 新加坡元 | 504.49 | 504.49 | 500.47 | 484.64 |
| 瑞士法郎 | 697.09 | 697.09 | 691.53 | 669.66 |

图 1-7-2　通过 show()函数显示的结果

## 1.8　实战项目　爬取实际外汇网站数据

**任务目标**

编写一个爬虫程序爬取银行网站外汇的汇率数据并将其存储到数据库。

### 1.8.1　解析网站的 HTML 代码

用 Chrome 浏览器访问招商银行网站，找到外汇汇率的其中一行数据（如美元），单击鼠标右键，在弹出的快捷菜单中选择"检查"命令，就可以看到对应的 HTML 代码，如图 1-8-1 所示。

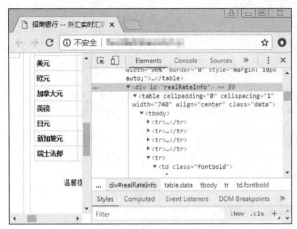

图 1-8-1　网站的 HTML 代码

从 HTML 代码中可以发现<div id="realRateInfo">…</div>中包含<table>…</table>，每行的<tr>…</tr>中就是不同外汇的汇率数据。复制相应的 HTML 代码，结果如下：

```
    <div id="realRateInfo">
      <table align="center" cellpadding="0" cellspacing="1" class="data" width="740">
       <tbody>
        <tr>
         <td class="head fontbold" width="70">
          交易币
         </td>
         <td class="head" width="65">
          交易币单位
         </td>
         <td class="head fontbold" width="55">
          基本币
         </td>
         <td class="head" width="65">
          现汇卖出价
         </td>
         <td class="head" width="65">
          现钞卖出价
         </td>
         <td class="head" width="65">
          现汇买入价
         </td>
         <td class="head" width="65">
          现钞买入价
         </td>
         <td class="head" width="65">
          时间
```

```
            </td>
            <td class="head">
                汇率走势图
            </td>
        </tr>
        ……
        </tbody>
    </table>
    <span class="tip">
        以上资料仅供参考，以办理业务时的实时汇率为准。
    </span>
</div>
```

很显然，这些外汇的汇率数据都嵌套在<table>…</table>的各个<td>…</td>中，找到这些<td>…</td>就可以得到想要的数据。

### 1.8.2 爬取网站外汇汇率数据

设计爬取过程如下：

（1）获取该网站的 HTML 字符串；

（2）用正则表达式匹配<div id="realRateInfo">和</div>，取出它们中间部分的字符串；

（3）匹配<tr>与</tr>，取出它们中间的字符串并命名为 tds；

（4）在 tds 中匹配<td>与</td>，取出各个<td>…</td>中的数据，并将数据存储到数据库。

值得注意的是，匹配<td>与</td>没有那么简单，因为各个<td>中还有一些属性，如<td class="numberright">等，不能简单地使用 r"<td>"的正则表达式去匹配 "td"。

数据包含在 HTML 代码的<td>…</td>中，为了爬取各个<td>…</td>中的数据，我们设计一个匹配函数 match()。

```
def match(t,s):
    m=re.search(r"<"+t,s)
    if m:
        a=m.start()
        m=re.search(r">",s[a:])
        if m:
            b=a+m.end()
            return {"start":a,"end":b}
    return None
```

这个函数从字符串 s 中匹配一个元素 t，例如，match("td",s)是在字符串 s 中匹配一个<td …>，它返回<td …>的开始位置与结束位置，无论这个<td …>中包含什么属性。

函数先使用语句 m=re.search(r"<"+t,s)匹配"<td"，找到它的开始位置 a=m.start()，然后在 s[a:]中匹配 ">"。

```
m=re.search(r">",s[a:])
```

那么 b=a+m.end()就是<td …>的结束位置，无论<td …>中有什么属性。通过 match("td",s)与 match("/td",s)语句就可以找<td …>…</td>了。

例如：

```
import re
def match(t,s):
```

```
            m=re.search(r"<"+t,s)
            if m:
                a=m.start()
                m=re.search(r">",s[a:])
                if m:
                    b=a+m.end()
                    return {"start":a,"end":b}
            return None
html='''
<tr>
    <td class="fontbold">美元</td>
    <td align="center">100</td>
    <td align="center" class="fontbold">人民币</td>
    <td class="numberright">686.94</td>
    <td class="numberright">686.94</td>
    <td class="numberright">684.00</td>
    <td class="numberright">678.38</td>
    <td align="center">13:40:36</td>
</tr>
'''
while True:
    m=match("td",html)
    n=match("/td",html)
    if m and n:
        u=m["end"]
        v=n["start"]
        s=html[u:v].strip("").strip("\n")
        print(s)
        html=html[n["end"]:]
```

程序结果:

```
美元
100
人民币
686.94
686.94
684.00
678.38
13:40:36
```

结果表明，match()函数能很好地匹配<td …>，无论<td …>中包含什么属性。

### 1.8.3 设计存储数据库

设计一个 SQLite3 数据库 rates.db，它有一张 rates 表，其结构如表 1-8-1 所示。

表 1-8-1　rates 表结构

| 字段名称 | 类型 | 说明 |
| --- | --- | --- |
| Currency | varchar(256) | 外汇名称（关键字） |
| TSP | float | 现汇卖出价 |

续表

| 字段名称 | 类型 | 说明 |
| --- | --- | --- |
| CSP | float | 现钞卖出价 |
| TBP | float | 现汇买入价 |
| CBP | float | 现钞买入价 |
| Time | varchar(256) | 时间 |

### 1.8.4 编写爬虫程序

爬虫程序采用两层循环，外层循环匹配<tr>…</tr>，内层循环匹配<td …>…</td>。爬虫程序如下：

```python
import urllib.request
import re
import sqlite3

class MySpider:
    def openDB(self):
        #初始化数据库，创建数据库 rates.db 与一张空表 rates
        self.con=sqlite3.connect("rates.db")
        self.cursor=self.con.cursor()
        try:
            self.cursor.execute("drop table rates")
        except:
            pass
        sql="create table rates (Currency varchar(256) primary key,TSP float,CSP float, TBP float, CBP float,Time varchar(256))"
        self.cursor.execute(sql)

    def closeDB(self):
        #关闭数据库
        self.con.commit()
        self.con.close()

    def insertDB(self,Currency,TSP,CSP,TBP,CBP,Time):
        #将记录插入数据库
        try:
            sql="insert  into  rates  (Currency,TSP,CSP,TBP,CBP,Time) values (?,?,?,?,?,?)"
            self.cursor.execute(sql,[Currency,TSP,CSP,TBP,CBP,Time])
        except Exception as err:
            print(err)

    def show(self):
        #显示函数
        self.cursor.execute("select Currency,TSP,CSP,TBP,CBP,Time from rates")
        rows=self.cursor.fetchall()
        print("%-18s%-12s%-12s%-12s%-12s%-12s"
```

```python
% ("Currency","TSP","CSP","TBP","CBP","Time"))
            for row in rows:
                print("%-18s%-12.2f%-12.2f%-12.2f%-12.2f%-12s" % (row[0],
row[1],row[2],row[3],row[4],row[5]))

    def match(self,t, s):
        #匹配函数
        m = re.search(r"<" + t, s)
        if m:
            a = m.start()
            m = re.search(r">", s[a:])
            if m:
                b = a + m.end()
                return {"start": a, "end": b}
        return None

    def spider(self,url):
        #爬虫函数
        try:
            resp = urllib.request.urlopen(url)
            data = resp.read()
            html = data.decode()
            m = re.search(r'<div id="realRateInfo">', html)
            html = html[m.end():]
            m = re.search(r'</div>', html)
            # 取出<div id="realRateInfo">…</div>部分
            html = html[:m.start()]
            i=0
            while True:
                p = self.match("tr", html)
                q = self.match("/tr", html)
                if p and q:
                    i=i+1
                    a = p["end"]
                    b = q["start"]
                    tds = html[a:b]
                    row=[]
                    count = 0
                    while True:
                        m = self.match("td", tds)
                        n = self.match("/td", tds)
                        if m and n:
                            u = m["end"]
                            v = n["start"]
                            count += 1
                            if count <= 8:
                                row.append(tds[u:v].strip())
                            tds = tds[n["end"]:]
                        else:
                            # 匹配不到<td ...>…</td>,退出内层循环
                            break
                    if i>=2 and len(row)==8:
```

```
                                    Currency = row[0]
                                    TSP = float(row[3])
                                    CSP = float(row[4])
                                    TBP = float(row[5])
                                    CBP = float(row[6])
                                    Time=row[7]
                                    self.insertDB(Currency,TSP,CSP,TBP,
CBP,Time)
                            html = html[q["end"]:]
                        else:
                            # 匹配不到<tr>…</tr>，退出外层循环
                            break
            except Exception as err:
                print(err)

    def process(self):
        #爬取过程函数
        self.openDB()
        self.spider("http://fx.cmbchina.com/hq/")
        self.show()
        self.closeDB()

#主程序
spider=MySpider()
spider.process()
```

程序说明如下。

程序的主要函数是 spider()函数，程序先获取网站的 HTML 代码，然后使用：

```
m=re.search(r'<div id="realRateInfo">',html)
html=html[m.end():]
m=re.search(r'</div>',html)
#取出<div id="realRateInfo">…</div>部分
html=html[:m.start()]
```

匹配到<div id="realRateInfo">…</div>，取出它们中间的部分，这部分字符串包含一个表格，表格中有很多<tr>…</tr>，使用外层循环匹配<tr>…</tr>。

```
p = match("tr", html)
q = match("/tr", html)
a=p["end"]
b=q["start"]
tds=html[a:b]
```

取出<tr>…</tr>的中间部分得到 tds，再使用内层循环匹配每个<td …>…</td>。

```
m = match("td", tds)
n = match("/td", tds)
```

tds 已经是包含<td>标签的字符串，从 tds 取出的 tds[m["end"]:n["start"]]部分就是所需的数据。

### 1.8.5　执行爬虫程序

执行爬虫程序，爬取网站的外汇汇率数据并将其存储到数据库中。爬取的部分结果如图 1-8-2 所示（数据与爬取时间有关）。

# 项目 ① 爬取外汇网站数据

| Currency | TSP | CSP | TBP | CBP | Time |
|---|---|---|---|---|---|
| 新西兰元 | 461.05 | 461.05 | 457.37 | 442.91 | 15:52:20 |
| 澳大利亚元 | 476.54 | 476.54 | 472.74 | 457.79 | 15:52:20 |
| 美元 | 672.84 | 672.84 | 669.95 | 664.45 | 15:52:20 |
| 欧元 | 759.22 | 759.22 | 753.18 | 729.35 | 15:52:20 |
| 加拿大元 | 502.75 | 502.75 | 498.75 | 482.97 | 15:52:20 |
| 英镑 | 891.76 | 891.76 | 884.66 | 856.68 | 15:52:20 |
| 日元 | 6.05 | 6.05 | 6.00 | 5.81 | 15:52:20 |
| 新加坡元 | 496.94 | 496.94 | 492.98 | 477.39 | 15:52:20 |
| 瑞士法郎 | 667.11 | 667.11 | 661.79 | 640.86 | 15:52:20 |

图 1-8-2　爬取的部分结果

## 项目总结

本项目涉及一个外汇网站，使用正则表达式对网站的网页进行解析，得到所需数据，实现了爬取网站外汇汇率数据的爬虫程序。

正则表达式匹配字符串的功能虽然强大（用它来解析一个简单的网页没有问题），但是使用它来解析复杂的网页比较困难，因为网站的网页结构不是简单的字符串，各个数据之间的关系实际上是树状结构关系，在后面的项目中将讲解更加高效的解析方法。

## 练习1

1. Flask 是一个比较简单的 Python Web 开发框架，你知道 Python 还有哪些比较流行的 Web 开发框架吗？说说它们的主要区别。

2. 使用 GET 方法提交数据与使用 POST 方法提交数据有什么不同？在 Flask 中如何获取提交的数据？

3. 使用 Flask 编写一个 Web 程序，它接收如下代码提交的 user_name 与 user_pass 数据。

```
<form name="frm" method="post" action="/login">
<input type="text" name="user_name" >
<input type="password" name="user_pass">
<input type="submit" value="Login" >
</form>
```

4. 说出下面正则表达式匹配的字符串分别是什么。

（1） r"\w+\s"

（2） r"\w+\b"

（3） r"\d+-\d+"

（4） r"\w+@(\w+\.)+\w+"

（5） r"(b|cd)ef"

5. 使用正则表达式匹配 HTML 代码中所有形如<img src="http:// …/文件名.jpg">的 JPG 图像文件，找出一个网站中所有这样表示的 JPG 图像文件，并下载这些图像文件。

# 项目 ② 爬取名言网站数据

拓展阅读

名人名言 50 句

爬虫程序的重要环节是在 HTML 代码中查找数据，查找的方法有很多，BeautifulSoup 程序包就是其中常用的一种，它是一个高效的解析与爬取程序包。本项目将学习使用 BeautifulSoup 来解析名人名言网站的 HTML 代码，爬取名人名言数据。

本项目使用的网站中有很多名人名句，我们可以编写爬虫程序爬取这些名句并做成小册子，认真阅读与思考，随时鞭策自己，激励自己。有志者，事竟成，我们只有胸怀大志，才能成为对社会有用之人。

2-1-A 知识讲解　　2-1-B 操作演练

## 2.1 项目任务

Quotes to Scrape 网站（http://quotes.toscrape.com/）是一个名言网站，进入该网站可以看到很多名人的名言，如图 2-1-1 所示。本项目介绍使用 BeautifulSoup 爬取数据的方法，并利用此方法爬取这个网站中所有名人的名言，将其存储到数据库。

在爬取这些数据之前，先练习爬取模拟名言网站的数据。设计一个类似的模拟名言网站，如图 2-1-2 所示，在掌握了基本的爬取技术后，再来爬取实际名言网站的数据。

图 2-1-1　实际名言网站

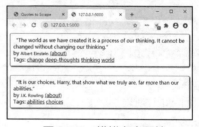
图 2-1-2　模拟名言网站

## 2.2 使用 BeautifulSoup 装载 HTML 文档

2-2-A 知识讲解　　2-2-B 操作演练

### 任务目标

创建一个名言网站以展示名人的名言，安装 BeautifulSoup 程序包，编写一个程序以获取该网站的 HTML 代码，使用 BeautifulSoup 装载 HTML 文档并按文档的树结构形式输出文档。在本节中，我们主要学习使用 BeautifulSoup 装载文档的方法。

# 项目 ❷  爬取名言网站数据

## 2.2.1  创建模拟名言网站

### 1. 创建网站模板

创建项目文件夹 project2，在它下面创建 templates 子文件夹，并在此子文件夹中创建一个文件 quotes.html，内容如下：

```html
        <style>
            .quote {
            padding: 10px;
            margin-bottom: 30px;
            border: 1px solid #333333;
            border-radius: 5px;
            box-shadow: 2px 2px 3px #333333;
            }
        </style>
        <div>
            <div class="quote" itemscope itemtype="http://schema.org/CreativeWork">
                <span class="text" itemprop="text">"The world as we have created it is a process of our thinking. It cannot be changed without changing our thinking."</span>
                <br><span>by <small class="author" itemprop="author">Albert Einstein</small>
                <a href="/author/Albert-Einstein">(about)</a>
                </span><br>
                <div class="tags">
                    Tags:
                    <meta class="keywords" itemprop="keywords" content="change,deep-thoughts,thinking,world" /   >
                        <a class="tag" href="/tag/change/page/1/">change</a>
                        <a class="tag" href="/tag/deep-thoughts/page/1/">deep-thoughts</a>
                        <a class="tag" href="/tag/thinking/page/1/">thinking</a>
                        <a class="tag" href="/tag/world/page/1/">world</a>
                </div>
            </div>
            <p></p>
            <div class="quote" itemscope itemtype="http://schema.org/CreativeWork">
                <span class="text" itemprop="text">"It is our choices, Harry, that show what we truly are, far more than our abilities."</span>
                <br><span>by <small class="author" itemprop="author">J.K. Rowling</small>
                <a href="/author/J-K-Rowling">(about)</a>
                </span><br>
                <div class="tags">
                    Tags:
                    <meta class="keywords" itemprop="keywords" content="abilities,choices" /   >
                        <a class="tag" href="/tag/abilities/page/1/">abilities</a>
                        <a class="tag" href="/tag/choices/page/1/">choices</a>
                </div>
```

31

```
            </div>
    </div>
```

### 2. 创建网站服务器程序

在 project2 中编写服务器程序 server.py，它的作用是返回 templates 中的 quotes.html 文件，程序如下：

```
import flask
app=flask.Flask(__name__)
@app.route("/")
def index():
    return flask.render_template("quotes.html")
app.run()
```

运行服务器程序后，用浏览器访问"http://127.0.0.1:5000"，就会看到图 2-1-2 所示的模拟名言网站。

## 2.2.2 安装 BeautifulSoup 程序包

HTML 元素的查找工具很多，BeautifulSoup 是其中十分流行的、功能非常强大的查找工具之一。BeautifulSoup 是第三方的工具，它包含在一个名称为 bs4 的程序包中，需要另外安装。其安装也很简单，在命令行窗口中进入 Python 的安装目录（如 C:\Python36），再进入 Scripts 子目录，找到 pip 程序，执行以下语句：

```
pip install bs4
```

安装完成后，执行下面的命令，安装 lxml 程序包：

```
pip install lxml
```

成功安装后就可以在 Python 的命令行窗口中执行语句：

```
from bs4 import BeautifulSoup
```

如果这条语句执行时没有报错，就说明安装成功了。

## 2.2.3 装载 HTML 文档

如果 html 是一个 HTML 文档，通过：

```
from bs4 import BeautifulSoup
soup=BequtifulSoup(html,"lxml")
```

就可以创建一个名称为 soup 的 BeautifulSoup 对象，其中，html 是一个 HTML 文档，"lxml"是一个参数，表示创建的是一个通过"lxml"解析器解析的文档。然后通过调用 soup.prettify() 函数可以获取 HTML 的文档树结构。

编写下列程序：

```
import urllib.request
from bs4 import BeautifulSoup
try:
    resp=urllib.request.urlopen("http://127.0.0.1:5000")
    html=resp.read().decode()
    soup=BeautifulSoup(html,"lxml")
    print(soup.prettify())
except Exception as err:
    print(err)
```

这个程序的功能很简单，就是访问模拟名言网站获取 html 字符串，创建一个 BeautifulSoup

对象 soup，再使用 soup.prettify()函数输出文档的树结构。程序运行的结果如下：

```html
<html>
 <head>
  <style>
   .quote {
     padding: 10px;
     margin-bottom: 30px;
     border: 1px solid #333333;
     border-radius: 5px;
     box-shadow: 2px 2px 3px #333333;
   }
  </style>
 </head>
 <body>
  <div>
   <div class="quote" itemscope="" itemtype="http://schema.org/CreativeWork">
    <span class="text" itemprop="text">
     "The world as we have created it is a process of our thinking. It cannot be changed without changing our thinking."
    </span>
    <br/>
    <span>
     by
     <small class="author" itemprop="author">
      Albert Einstein
     </small>
     <a href="/author/Albert-Einstein">
      (about)
     </a>
    </span>
    <br/>
    <div class="tags">
     Tags:
     <meta class="keywords" content="change,deep-thoughts,thinking,world" itemprop="keywords"/>
     <a class="tag" href="/tag/change/page/1/">
      change
     </a>
     <a class="tag" href="/tag/deep-thoughts/page/1/">
      deep-thoughts
     </a>
     <a class="tag" href="/tag/thinking/page/1/">
      thinking
     </a>
     <a class="tag" href="/tag/world/page/1/">
      world
     </a>
    </div>
   </div>
   <p>
   </p>
```

```
            <div class="quote" itemscope="" itemtype="http://schema.org/
CreativeWork">
                <span class="text" itemprop="text">
                "It is our choices, Harry, that show what we truly are, far more than
our abilities."
                </span>
                <br/>
                <span>
                 by
                <small class="author" itemprop="author">
                 J.K. Rowling
                </small>
                <a href="/author/J-K-Rowling">
                 (about)
                </a>
                </span>
                <br/>
                <div class="tags">
                Tags:
                <meta class="keywords" content="abilities,choices" itemprop=
"keywords"/>
                <a class="tag" href="/tag/abilities/page/1/">
                 abilities
                </a>
                <a class="tag" href="/tag/choices/page/1/">
                 choices
                </a>
                </div>
            </div>
        </div>
    </body>
</html>
```

这个结构类似于可扩展标记语言（eXtensible Markup Language，XML）文档结构。实际上，HTML 文档树严格来说就是 XML 文档结构，读者必须充分熟悉这种结构，它是爬取数据的关键。

## 2.3 使用 BeautifulSoup 查找 HTML 元素

**任务目标**

使用 BeautifulSoup 装载 HTML 文档的过程实际上是在内存中建立一棵文档树的过程，这棵树的节点就是各个 HTML 元素，数据就嵌在这些节点中。这棵树被建立后，我们可以沿树的任意一条路径去查找各个节点的数据。在本节中，我们主要学习从这个文档树中解析或者爬取所要的数据的方法。

### 2.3.1 使用 find()函数查找

查找文档的元素是爬取网页数据的重要手段。前面我们使用了正则表达式解析和爬取所要的数据，但这种方法只适合简单的 HTML 文档，如果 HTML 文档比较复杂，就要用

## 项目 ❷ 爬取名言网站数据

到其他的解析工具了，其中 BeautifulSoup 就是一个功能强大的解析工具。BeautifulSoup 提供了一系列查找元素的方法，例如，使用 find() 函数与 find_all() 函数查找元素。

find() 函数的基本用法如下：

```
find(self, name=None, attrs={},recursive=True)
```

self 表明它是一个类成员函数。

name 表示要查找的元素名称，默认是 None，如果不提供就查找所有元素。

attrs 表示元素的属性，它是一个字典，默认为空，如果提供就查找有这个指定属性的元素。

recursive=True 表示从当前元素开始，一直深入它的子树，进行全树范围查找。如果 recursive=False，则只在元素下一层的直接子节点中查找。

如果 find() 函数找到了元素，就返回一个 bs4 程序包中定义的 bs4.element.Tag 对象；如果没有找到，就返回 None。如果网站中有多个满足条件的元素，那么 find() 函数只返回第一个查找到的元素。

**例 2-3-1**：查找名言网站中的 \<span\> 元素。

```
import urllib.request
from bs4 import BeautifulSoup
resp=urllib.request.urlopen("http://127.0.0.1:5000")
html=resp.read().decode()
soup=BeautifulSoup(html,"lxml")
e=soup.find("span")
print(type(e))
print(e)
```

结果：

```
<class 'bs4.element.Tag'>
<span class="text" itemprop="text">"The world as we have created it is a process of our thinking. It cannot be changed without changing our thinking."</span>
```

由此可见，使用 soup.find("span") 找到了第一个 \<span\> 元素，返回了一个 bs4.element.Tag 对象，该对象的字符串显示的就是 \<span\> 元素。

**例 2-3-2**：查找名言网站中的 \<a\> 元素。

```
import urllib.request
from bs4 import BeautifulSoup
resp=urllib.request.urlopen("http://127.0.0.1:5000")
html=resp.read().decode()
soup=BeautifulSoup(html,"lxml")
print(soup.find("a"))
```

结果：

```
<a href="/author/Albert-Einstein">(about)</a>
```

网站中有多个 \<a\> 元素，find() 函数只负责找到第一个 \<a\> 元素。

**例 2-3-3**：查找名言网站中的 \<h4\> 元素。

```
import urllib.request
from bs4 import BeautifulSoup
resp=urllib.request.urlopen("http://127.0.0.1:5000")
html=resp.read().decode()
soup=BeautifulSoup(html,"lxml")
print(soup.find("h4"))
```

结果：

```
None
```

由此可见，没有找到<h4>元素，返回 None。

**例 2-3-4**：查找<a class="tag">元素。

网站中有多个<a>元素，通过 find()函数的 attrs 参数限定 find()函数查找 class="tag"的<a>元素，即可以缩小搜索范围，使得 find()函数快速找到指定的元素。

```python
import urllib.request
from bs4 import BeautifulSoup
resp=urllib.request.urlopen("http://127.0.0.1:5000")
html=resp.read().decode()
soup=BeautifulSoup(html,"lxml")
print(soup.find("a",attrs={"class":"tag"}))
```

结果：

```html
<a class="tag" href="/tag/change/page/1/">change</a>
```

找到了第一个<a class="tag">元素。

**例 2-3-5**：查找第一条名言中的第一个<div class="tags">中的第一个<a>元素。

首先使用 soup.find("div",attrs={"class":"tags"})找到第一个<div class="tags">元素，然后查找这个元素中的第一个<a>元素。

```python
import urllib.request
from bs4 import BeautifulSoup
resp=urllib.request.urlopen("http://127.0.0.1:5000")
html=resp.read().decode()
soup=BeautifulSoup(html,"lxml")
div=soup.find("div",attrs={"class":"tags"})
print(div.find("a"))
```

结果：

```html
<a class="tag" href="/tag/change/page/1/">change</a>
```

实际上，soup 是一个 bs4.element.Tag 对象，它是文档树的根，因此 div=soup.find("div",attrs={"class":"tags"})表示在全文档查找第一个<div class="tags">元素。而 div 也是一个 bs4.element.Tag 对象，因此 div.find("a")表示在<div class="tags">元素下面的子树中查找<a>元素，即查找它下面的第一个<a>元素。

最后两条语句可以连在一起写成：

```python
print(soup.find("div",attrs={"class":"tags"}).find("a"))
```

当然也可以先找到<div class="quote">，然后继续往下找，即：

```python
print(soup.find("div",attrs={"class":"quote"}).find("div",attrs={"class":"tags"}).find("a"))
```

**例 2-3-6**：查找第一条名言中的第一个<small>元素。

```python
import urllib.request
from bs4 import BeautifulSoup
resp=urllib.request.urlopen("http://127.0.0.1:5000")
html=resp.read().decode()
soup=BeautifulSoup(html,"lxml")
print(soup.find("small"))
```

结果：

```html
<small class="author" itemprop="author">Albert Einstein</small>
```

注意，如果使用：
```
print(soup.find("small",recursive=False))
```
那么结果是 None，因为<small>是在<div class="quote">元素下面的<span>元素中的，使用 recursive=False 表示从 soup 顶层开始不递归地向下层搜索，因此搜索不到<small>元素。同样，使用下面的语句也搜索不到<small>元素。
```
print(soup.find("div",attrs={"class":"quote"}).find("small",recursive=False))
```
但是使用下面的两条语句都能搜索到<small>元素。
```
print(soup.find("div",attrs={"class":"quote"}).find("small",recursive=True))
print(soup.find("div",attrs={"class":"quote"}).find("small"))
```
由此可见，recursive 参数默认为 True，表示搜索元素时会递归地往下层搜索。如果将其设置为 False，就只在第一层进行搜索，不会往下层递归搜索。

### 2.3.2 查找元素属性与文本

find()函数找到一个 HTML 元素，返回 bs4.element.Tag 对象的 tag，然后进行如下操作：

（1）使用 tag[attrName]获取名称为 attrName 的属性，如果这个元素没有 attrName 属性就抛出异常；

（2）使用 tag.text 获取 tag 元素下面的文本字符串，如果 tag 元素下面有其他的子节点元素，那么 tag.text 是所有子节点元素的文本的组合。

例 2-3-7：查找名言的文本。
```
import urllib.request
from bs4 import BeautifulSoup
resp=urllib.request.urlopen("http://127.0.0.1:5000")
html=resp.read().decode()
soup=BeautifulSoup(html,"lxml")
print(soup.find("span").text)
```
结果：
```
"The world as we have created it is a process of our thinking. It cannot be changed without changing our thinking."
```

例 2-3-8：查找名言对应的名人与链接。
```
import urllib.request
from bs4 import BeautifulSoup
resp=urllib.request.urlopen("http://127.0.0.1:5000")
html=resp.read().decode()
soup=BeautifulSoup(html,"lxml")
print(soup.find("small").text)
print(soup.find("a")["href"])
```
结果：
```
Albert-Einstein
/author/Albert-Einstein
```

例 2-3-9：查找<div class="tags">元素的文本。
```
import urllib.request
from bs4 import BeautifulSoup
resp=urllib.request.urlopen("http://127.0.0.1:5000")
```

```
html=resp.read().decode()
soup=BeautifulSoup(html,"lxml")
print(soup.find("div",attrs={"class":"tags"}).text)
```

结果：

```
            Tags:
change
deep-thoughts
thinking
world
```

值得注意的是，\<div class="tags"\>下除了直接的 Tags 文本外还包含多个\<a\>元素，将这些元素的文本全部集中起来作为 soup.find("div",attrs={"class":"tags"}).text 的结果。

### 2.3.3 使用 find_all()函数查找

find()函数只负责查找第一个满足条件的元素，而 find_all()函数负责查找所有满足条件的元素。find_all()函数的基本用法如下。

```
find_all(self, name=None, attrs={},recursive=True)
```

self 表明它是一个类成员函数。

name 表示要查找的元素名称，默认是 None，如果不提供就查找所有的元素。

attrs 表示元素的属性，它是一个字典，默认是空，如果提供就查找有这个指定属性的元素。

find_all()函数若查找成功，则返回找到的所有元素的 bs4.element.Tag 对象列表，如果找不到就返回空列表。

recursive=True 表示从当前元素开始，一直深入它的子树，进行全树范围查找。如果 recursive=False，则只在元素下一层的直接子节点中查找。

例 2-3-10：查找所有\<a\>元素的文本。

```
import urllib.request
from bs4 import BeautifulSoup
resp=urllib.request.urlopen("http://127.0.0.1:5000")
html=resp.read().decode()
soup=BeautifulSoup(html,"lxml")
links=soup.find_all("a")
for link in links:
    print(link.text)
```

结果：

```
(about)
change
deep-thoughts
thinking
world
(about)
abilities
choices
```

其中，soup.find_all("a")用于查找所有\<a\>元素，一共有 8 个，组成了一个列表 links，使用循环得到每个 link，再使用 link.text 得到每个 link 的文本值。

## 项目 ❷　爬取名言网站数据

**例 2-3-11**：查找所有名言。

```
import urllib.request
from bs4 import BeautifulSoup
resp=urllib.request.urlopen("http://127.0.0.1:5000")
html=resp.read().decode()
soup=BeautifulSoup(html,"lxml")
divs=soup.find_all("div",attrs={"class":"quote"})
for div in divs:
    print(div.find("span").text)
```

结果：

```
"The world as we have created it is a process of our thinking. It cannot be changed without changing our thinking."
"It is our choices, Harry, that show what we truly are, far more than our abilities."
```

其中，soup.find_all("div",attrs={"class":"quote"})用于查找所有<div class="quote">元素，一共有两个，组成了 divs 列表，再次循环 divs 列表的每个元素<div>，使用 div.find("span").text 找到了名言的文本。

**例 2-3-12**：查找最后一条名言与对应的名人。

```
import urllib.request
from bs4 import BeautifulSoup
resp=urllib.request.urlopen("http://127.0.0.1:5000")
html=resp.read().decode()
soup=BeautifulSoup(html,"lxml")
div=soup.find_all("div",attrs={"class":"quote"})[-1]
print(div.find("span").text)
print(div.find("small").text)
```

结果：

```
"It is our choices, Harry, that show what we truly are, far more than our abilities."
J.K. Rowling
```

其中，div=soup.find_all("div",attrs={"class":"quote"})[-1]用于查找最后一个<div class="quote">元素。

**例 2-3-13**：查找所有<div class="tags">元素的最后一个<a>元素的链接。

```
import urllib.request
from bs4 import BeautifulSoup
resp=urllib.request.urlopen("http://127.0.0.1:5000")
html=resp.read().decode()
soup=BeautifulSoup(html,"lxml")
divs=soup.find_all("div",attrs={"class":"quote"})
for div in divs:
    links=div.find("div",attrs={"class":"tags"}).find_all("a")
    print(links[-1]["href"])
```

结果：

```
/tag/world/page/1/
/tag/choices/page/1/
```

其中，links=div.find("div",attrs={"class":"tags"}).find_all("a")用于查找<div class="tags">元素中的所有<a>元素。

39

例 2-3-14：查找所有<a class="tag">元素的链接。

```
import urllib.request
from bs4 import BeautifulSoup
resp=urllib.request.urlopen("http://127.0.0.1:5000")
html=resp.read().decode()
soup=BeautifulSoup(html,"lxml")
divs=soup.find_all("div",attrs={"class":"quote"})
for div in divs:
    print(div.find("span").text)
    links=div.find_all("a",attrs={"class":"tag"})
    for link in links:
        print(link.["href"])
```

结果：

```
"The world as we have created it is a process of our thinking. It cannot be changed without changing our thinking."
/tag/change/page/1/
/tag/deep-thoughts/page/1/
/tag/thinking/page/1/
/tag/world/page/1/
"It is our choices, Harry, that show what we truly are, far more than our abilities."
/tag/abilities/page/1/
/tag/choices/page/1/
```

### 2.3.4 使用高级查找

一般情况下，find()或者 find_all()函数能满足人们的查找需要，如果不能满足需要，那么可以设计一个查找函数来进行查找。设计一个查找函数，如 myFilter()，则可以使用 find(myFilter)或者 find_all(myFilter)进行查找。myFilter()函数的结构如下。

```
def myFilter(tag):
    #tag 是传递的 bs4.element.Tag 对象
    #返回 True 表示查找成功，返回 False 表示查找失败
```

例 2-3-15：使用高级查找函数。

```
import urllib.request
from bs4 import BeautifulSoup
def myFilter(tag):
    print(tag.name,end=",")
resp=urllib.request.urlopen("http://127.0.0.1:5000")
html=resp.read().decode()
soup=BeautifulSoup(html,"lxml")
soup.find("div",attrs={"class":"quote"}).find(myFilter)
```

结果：

```
span,br,span,small,a,br,div,meta,a,a,a,a,
```

由此可见，使用 soup.find("div",attrs={"class":"quote"})得到第一个<div class="quote">元素，它调用 find(myFilter)时使用高级查找函数，会把该元素下的每个元素传递给 tag 参数，因此 span、br、span、small、a、br、div、meta、a、a、a、a 等元素被传递给了 myFilter()中的 tag 参数。

例 2-3-16：查找文本值为 thinking 的<a>元素。

```
import urllib.request
from bs4 import BeautifulSoup
def myFilter(tag):
    if tag.name=="a" and tag.text=="thinking":
        return True
    return False
resp=urllib.request.urlopen("http://127.0.0.1:5000")
html=resp.read().decode()
soup=BeautifulSoup(html,"lxml")
print(soup.find("div",attrs={"class":"quote"}).find(myFilter))
```

结果：

```
<a class="tag" href="/tag/thinking/page/1/">thinking</a>
```

程序在执行 find(myFilter)时会传递每一个元素到 tag 参数，tag 是一个 bs4.element.Tag 对象，通过 tag.name 得到该元素的名称，通过 tag.text 得到该元素的文本。因此一旦发现某元素的 tag.name 为"a"且 tag.text 为"thinking"时，这个元素就是要寻找的<a>元素。

例 2-3-17：查找所有<a class="tag">元素的文本。

```
import urllib.request
from bs4 import BeautifulSoup
def myFilter(tag):
    if tag.name=="a" and tag.has_attr("class") and tag["class"]==["tag"]:
        return True
    return False
resp=urllib.request.urlopen("http://127.0.0.1:5000")
html=resp.read().decode()
soup=BeautifulSoup(html,"lxml")
links=soup.find_all(myFilter)
for link in links:
    print(link)
```

结果：

```
<a class="tag" href="/tag/change/page/1/">change</a>
<a class="tag" href="/tag/deep-thoughts/page/1/">deep-thoughts</a>
<a class="tag" href="/tag/thinking/page/1/">thinking</a>
<a class="tag" href="/tag/world/page/1/">world</a>
<a class="tag" href="/tag/abilities/page/1/">abilities</a>
<a class="tag" href="/tag/choices/page/1/">choices</a>
```

其中，tag.name=="a"要求 tag 为<a>的元素，tag.has_attr("class")要求 tag 元素有 class 属性，而 tag["class"]==["tag"]要求 class 值包含"tag"。

**注意**：class 的值都要写成列表形式，因为 class 的值可能有多个，如果写成 tag["class"]=="tag"，则找不到<a class="tag">元素。

另外，myFilter()函数中还必须写出 tag.has_attr("class")的条件，因为如果函数变成：

```
def myFilter(tag):
    if tag.name=="a" and tag["class"]==["tag"]:
        return True
    return False
```

那么执行时会出现错误，原因是有些<a>元素没有 class 属性，判断 tag["class"]==["tag"]时就会抛出异常。

例 2-3-18：查找所有 href 属性中包含 page 关键字的<a>元素。

```
import urllib.request
from bs4 import BeautifulSoup
def myFilter(tag):
    if tag.name=="a" and tag.has_attr("href") and tag["href"].find("page")>=0:
        return True
    return False
resp=urllib.request.urlopen("http://127.0.0.1:5000")
html=resp.read().decode()
soup=BeautifulSoup(html,"lxml")
links=soup.find_all(myFilter)
for link in links:
    print(link)
```

结果：

```
<a class="tag" href="/tag/change/page/1/">change</a>
<a class="tag" href="/tag/deep-thoughts/page/1/">deep-thoughts</a>
<a class="tag" href="/tag/thinking/page/1/">thinking</a>
<a class="tag" href="/tag/world/page/1/">world</a>
<a class="tag" href="/tag/abilities/page/1/">abilities</a>
<a class="tag" href="/tag/choices/page/1/">choices</a>
```

其中，tag["href"].find("page")>=0 要求 href 属性包含 "page" 关键字。

## 2.4 使用 BeautifulSoup 遍历文档元素

2-4-A 知识讲解　2-4-B 操作演练

 **任务目标**

通过 BeautifulSoup 不但可以在 HTML 文档中获取各种特征的元素节点数据，还可以遍历 HTML 文档树的各个节点，获取节点的父节点、子孙节点、兄弟节点等，这对于查找那些没有明显特征的元素节点十分有用。

### 2.4.1 获取元素节点的父节点

如果 tag 是一个 bs4.element.Tag 对象，那么可通过下面的语句获取 tag 的父节点：

```
tag.parent
```

一般来说，每个节点都有一个父节点，其中根节点<html>的父节点是名称为[document]的节点。[document]是 HTML 文档树的最高节点，[document]节点的父节点是 None。

例 2-4-1：找出网站中第一个<span>节点的所有父节点的名称。

```
import urllib.request
from bs4 import BeautifulSoup
resp=urllib.request.urlopen("http://127.0.0.1:5000")
html=resp.read().decode()
soup=BeautifulSoup(html,"lxml")
p=soup.find("span")
while p:
    print(p.name)
    p=p.parent
```

结果：
```
span
div
div
body
html
[document]
```
第一个<span>的父节点是<div class="quote">，该父节点的父节点是<div>，再上一级的父节点是<body>以及<html>，然后是[document]。

## 2.4.2 获取元素节点的直接子节点

如果 tag 是一个 bs4.element.Tag 对象，那么可通过下面的语句获取 tag 节点的所有直接子节点：

```
tag.children
```

子节点包括元素（element）、文本（text）等类型的节点。element 类型的节点是 bs4.element.Tag 对象，text 类型的节点是 bs4.element.NavigableString 对象。

**例 2-4-2**：找出第一个<div class="quote">下面的第二个<span>下面的所有直接子节点的类型与名称。

```
import urllib.request
from bs4 import BeautifulSoup
import bs4
resp=urllib.request.urlopen("http://127.0.0.1:5000")
html=resp.read().decode()
soup=BeautifulSoup(html,"lxml")
p=soup.find("div",attrs={"class":"quote"}).find_all("span")[1]
for q in p.children:
    print(type(q),q.name)
    if isinstance(q,bs4.element.NavigableString):
        print(q.string)
    else:
        print(q.text)
```

结果：
```
<class 'bs4.element.NavigableString'> None

by
<class 'bs4.element.Tag'> small
Albert Einstein
<class 'bs4.element.NavigableString'> None

<class 'bs4.element.Tag'> a
(about)
<class 'bs4.element.NavigableString'> None
```

值得注意的是，结果中包含一些空格与换行字符，它们的类型为 bs4.element.NavigableString，这种文本节点没有 name 属性值与 text 属性值，它们的文本使用 string 属性值获取。

**例 2-4-3**：查看网站中文本的直接子节点。

```
import urllib.request
from bs4 import BeautifulSoup
import bs4
def show(st):
```

```
    print(len(st),": ",end="")
    for s in st:
      print(hex(ord(s)),end=" ")
    print()
resp=urllib.request.urlopen("http://127.0.0.1:5000")
html=resp.read().decode()
soup=BeautifulSoup(html,"lxml")
p=soup.find("div",attrs={"class":"quote"}).find_all("span")[1]
for q in p.children:
    if isinstance(q,bs4.element.NavigableString):
        show(q)
```

结果：

```
3 : 0x62 0x79 0x20
1 : 0xa
1 : 0xa
```

由此可见,"by"后面有一个空格,结果对应的编码是"0x62 0x79 0x20",而文本中两个 bs4.element.NavigableString 类型的节点的值是换行符 "\n",编码是"0xa"。

**例 2-4-4**：查看第一个<div class="tags">元素中非 "\n" 的文本。

```
import urllib.request
from bs4 import BeautifulSoup
import bs4
resp=urllib.request.urlopen("http://127.0.0.1:5000")
html=resp.read().decode()
soup=BeautifulSoup(html,"lxml")
p=soup.find("div",attrs={"class":"tags"})
for q in p.children:
    if isinstance(q,bs4.element.NavigableString) and q.string!="\x0a":
        print(q.string.strip())
```

结果：

```
Tags:
```

### 2.4.3 获取元素节点的所有子孙节点

如果 tag 是一个 bs4.element.Tag 对象,那么可通过下面的语句获取 tag 节点的所有子孙节点：

```
tag.descendants
```

子孙节点包括 element、text 等类型的节点。

**例 2-4-5**：找出网站中第一个<div class="tags">节点的所有子孙节点的类型与名称。

```
import urllib.request
from bs4 import BeautifulSoup
import bs4
resp=urllib.request.urlopen("http://127.0.0.1:5000")
html=resp.read().decode()
soup=BeautifulSoup(html,"lxml")
p=soup.find("div",attrs={"class":"tags"})
for q in p.descendants:
    print(type(q),q.name)
    if isinstance(q,bs4.element.NavigableString):
        s=q.string.strip("\n")
        s=s.strip()
        if s!="":
```

```
            print(s)
    else:
        print(q.name,q.text)
```
结果：
```
<class 'bs4.element.NavigableString'> None
Tags:
<class 'bs4.element.Tag'> meta
meta
<class 'bs4.element.NavigableString'> None
<class 'bs4.element.Tag'> a
a change
<class 'bs4.element.NavigableString'> None
change
<class 'bs4.element.NavigableString'> None
<class 'bs4.element.Tag'> a
a deep-thoughts
<class 'bs4.element.NavigableString'> None
deep-thoughts
<class 'bs4.element.NavigableString'> None
<class 'bs4.element.Tag'> a
a thinking
<class 'bs4.element.NavigableString'> None
thinking
<class 'bs4.element.NavigableString'> None
<class 'bs4.element.Tag'> a
a world
<class 'bs4.element.NavigableString'> None
world
<class 'bs4.element.NavigableString'> None
```
由此可见，tag.descendants 获取的是所有的子孙节点。

### 2.4.4 获取元素节点的兄弟节点

如果 tag 是一个 bs4.element.Tag 对象，那么可通过下列语句来获取其后一个和前一个兄弟节点：

```
tag.next_sibling
tag.previous_sibling
```

其中，tag.next_sibling 是 tag 的邻近的后一个兄弟节点，tag.previous_sibling 是 tag 的邻近的前一个兄弟节点。兄弟节点可以是 element 类型的节点，也可以是 text 等类型的节点。

例 2-4-6：查找元素的兄弟节点。

```
import urllib.request
from bs4 import BeautifulSoup
import bs4
resp=urllib.request.urlopen("http://127.0.0.1:5000")
html=resp.read().decode()
soup=BeautifulSoup(html,"lxml")
tag=soup.find("div",attrs={"class":"quote"}).find("span")
p=tag.previous_sibling
while p and not isinstance(p,bs4.element.Tag):
    p=p.previous_sibling
```

```
    if p:
      print(p.name)
    else:
      print("None")
    p=tag.next_sibling
    while p and not isinstance(p,bs4.element.Tag):
      p=p.next_sibling
    if p:
      print(p.name)
    else:
      print("None")
```

结果:

```
None
br
```

其中，tag=soup.find("div",attrs={"class":"quote"}).find("span")用于查找第一个<span>，一般前后兄弟节点可能是一些空白字符或者换行字符。使用 while 循环：

```
    while p and not isinstance(p,bs4.element.Tag):
        p=p.previous_sibling
```

找到的前一个兄弟节点是一个 bs4.element.Tag 对象，发现它没有前序的兄弟节点。同样，使用：

```
    while p and not isinstance(p,bs4.element.Tag):
        p=p.next_sibling
```

找到的后一个兄弟节点是一个 bs4.element.Tag 对象，发现它是<br>元素。

例 2-4-7：查找第一个<div class="tags">元素下面所有并列的<a>元素。

```
import urllib.request
from bs4 import BeautifulSoup
import bs4
resp=urllib.request.urlopen("http://127.0.0.1:5000")
html=resp.read().decode()
soup=BeautifulSoup(html,"lxml")
p=soup.find("div",attrs={"class":"tags"}).find("a")
while p:
  if isinstance(p,bs4.element.Tag) and p.name=="a":
      print(p)
  p=p.next_sibling
```

结果:

```
<a class="tag" href="/tag/change/page/1/">change</a>
<a class="tag" href="/tag/deep-thoughts/page/1/">deep-thoughts</a>
<a class="tag" href="/tag/thinking/page/1/">thinking</a>
<a class="tag" href="/tag/world/page/1/">world</a>
```

程序先查找第一个<a>元素，然后继续查找它的下一个<a>兄弟元素，直到找到所有元素。

## 2.5 BeautifulSoup 支持使用 CSS 语法进行查找

2-5-A  知识讲解

2-5-B  操作演练

### 任务目标

BeautifulSoup 的 find()函数与 find_all()函数功能很强大，同时 BeautifulSoup 支持使用 CSS 语法的查找方法，这对熟悉 CSS 语法规则的读者而言是一个"福音"，读者可以通过 CSS 语法查找元素。

## 2.5.1 使用 CSS 语法查找

除了可以用 find()函数与 find_all()函数查找 HTML 文档树的节点元素外，还可以采用与 CSS 类似的语法来查找，规则是：

```
tag.select(css)
```

其中，tag 是一个 bs4.element.Tag 对象，即 HTML 中的一个 element 类型的节点，select()是它的查找方法，css 是类似 CSS 语法的一个字符串，一般结构如下：

```
[tagName][attName[=value]]
```

其中，[]中的内容是可选的。

tagName 表示元素名称，如果没有指定就查找所有元素。

attName 表示属性名称，value 是它对应的值，可以不指定属性，在指定了属性后也可以不指定值。

tag.select(css)返回一个 bs4.element.Tag 列表，其中只含有一个元素。

如果有多个元素，还可以使用:nth-of-type(n)来指定返回第 n 个元素，其中 n=1 表示指定返回第一个元素。

如果 attName 的值是 class，那么可以把"class=value"写成".value"。例如，<p class="xxx">可以写成 p[class="xxx"]，也可以写成 p.xxx。

如果 attName 的值是 id，那么可以把"id=value"写成"#value"。例如，<p id="xxx">可以写成 p[id="xxx"]，也可以写成 p.#xxx。

BeautifulSoup 除了 select()函数外，还有一个 select_one()函数。如果只查找一个元素，那么还可以使用：

```
tag.select_one(css)
```

查找第一个满足要求的元素，如果找不到指定的元素就返回 None。

例 2-5-1：使用 select()查找举例。

（1）查找文档中<li>节点下的所有<a>节点。
```
soup.select("li a")
```
（2）查找文档中属性 class="sighthotel"的<div>节点下的所有<span>节点。
```
soup.select("div[class='sighthotel'] span")
```
（3）查找文档中具有 class 属性的<li>节点下的所有<a>节点。
```
soup.select("li[class] a")
```
（4）查找<div>下面具有 title 属性的<li>节点。
```
soup.select("div li[title]")
```
（5）查找<body>下面的<head>下面的<title>节点。
```
soup.select("body head title")
```
（6）查找<body>下面所有具有 class 属性的节点。
```
soup.select("body [class] ")
```
（7）查找<body>下面所有具有 class 属性的节点下面的<a>节点。
```
soup.select("body [class] a")
```
（8）查找<body>下面所有具有 class 属性的节点下面的第二个<a>节点。
```
soup.select("body [class] a:nth-of-type(2)")
```

例 2-5-2：查找网站中<div class="quote">中的所有<span>下面的<a>节点。

```python
import urllib.request
from bs4 import BeautifulSoup
import bs4
resp=urllib.request.urlopen("http://127.0.0.1:5000")
html=resp.read().decode()
soup=BeautifulSoup(html,"lxml")
links=soup.select("div[class='quote'] span a")
for link in links:
    print(link)
```

结果：

```
<a href="/author/Albert-Einstein">(about)</a>
<a href="/author/J-K-Rowling">(about)</a>
```

其中，soup.select("div[class='quote'] span a")用于查找<div class="quote">中<span>元素下的所有<a>节点。

例 2-5-3：查找网站中<div class="quote">下面的第二个<span>。

```python
import urllib.request
from bs4 import BeautifulSoup
resp=urllib.request.urlopen("http://127.0.0.1:5000")
html=resp.read().decode()
soup=BeautifulSoup(html,"lxml")
spans=soup.select("div[class='quote'] span:nth-of-type(2)")
for span in spans:
    print(span)
```

结果：

```
<span>by <small class="author" itemprop="author">Albert Einstein</small>
<a href="/author/Albert-Einstein">(about)</a>
</span>
<span>by <small class="author" itemprop="author">J.K. Rowling</small>
<a href="/author/J-K-Rowling">(about)</a>
</span>
```

### 2.5.2 使用属性的语法规则

CSS 结构中的[attName=value]表示属性 attName 与 value 相等，也可以指定不等、包含等。具体语法规则如表 2-5-1 所示。

表 2-5-1 CSS 属性语法规则

| 选择器 | 描述 |
| --- | --- |
| [attName] | 用于选取带有指定属性的元素 |
| [attName=value] | 用于选取带有指定属性和值的元素 |
| [attName^=value] | 匹配属性值以指定值开头的每个元素 |
| [attName$=value] | 匹配属性值以指定值结尾的每个元素 |
| [attrName*=value] | 匹配属性值中包含指定值的每个元素 |

因此：

soup.select("a[href]")表示查找有 href 属性的<a>元素；

soup.select("a[href$='.html']")表示查找 href 以 ".html" 结尾的<a>节点；

soup.select("a[href^='http://']")表示查找 href 以 "http://" 开头的<a>节点；

soup.select("a[href*='example']")表示查找 href 的值中包含 "example" 字符串的<a>节点。

## 2.5.3 使用 select() 查找子孙节点

当 select(css) 中的 css 有多个节点时，节点元素之间用空格分开，表示查找子孙节点。例如，soup.select("div ul") 表示查找所有 <div> 节点下的所有子孙节点 <ul>。

例 2-5-4：查找子孙节点。

```
import urllib.request
from bs4 import BeautifulSoup
import bs4
resp=urllib.request.urlopen("http://127.0.0.1:5000")
html=resp.read().decode()
soup=BeautifulSoup(html,"lxml")
elems=soup.select("div[class='quote'] div[class='tags'] a:nth-of-type(2)")
for elem in elems:
    print(elem)
```

结果：

```
<a class="tag" href="/tag/deep-thoughts/page/1/">deep-thoughts</a>
<a class="tag" href="/tag/choices/page/1/">choices</a>
```

其中，soup.select("div[class='quote'] div[class='tags'] a:nth-of-type(2)") 用于查找 <div class="quote"> 下的 <div class='tags'> 下的第二个 <a> 节点。

## 2.5.4 使用 select() 查找直接子节点

当 select(css) 中的 css 有多个节点时，节点元素之间用 ">" 分开（注意，">" 的前后至少包含一个空格），表示查找直接子节点。例如，soup.select("div > p") 表示查找 <div> 节点下面的所有直接子节点 <p>，注意不包含子孙节点。

例 2-5-5：查找直接子节点。

```
import urllib.request
from bs4 import BeautifulSoup
resp=urllib.request.urlopen("http://127.0.0.1:5000")
html=resp.read().decode()
soup=BeautifulSoup(html,"lxml")
elems=soup.select("div[class='quote'] >a")
for elem in elems:
    print(elem)
```

结果什么都没有找到，因为 <div class="quote"> 下面的直接子节点元素中没有 <a> 节点。但是使用：

```
elems=soup.select("div[class='quote']> span >a")
for elem in elems:
    print(elem)
```

可以找到两个 <a> 节点。

```
<a href="/author/Albert-Einstein">(about)</a>
<a href="/author/J-K-Rowling">(about)</a>
```

## 2.5.5 使用 select() 查找兄弟节点

在 select() 中用 "~" 连接两个节点时表示查找前一个节点后面的所有同级别的兄弟节点（注意，"~" 前后至少有一个空格）。例如，soup.select("div ~ p") 表示查找 <div> 后面的所有同级别的<p>兄弟节点。

在 select() 中用 "+" 连接两个节点时表示查找前一个节点后面的第一个同级别的兄弟节点（注意，"+" 前后至少有一个空格）。

**例 2-5-6**：查找所有兄弟节点。

```
import urllib.request
from bs4 import BeautifulSoup
import bs4
resp=urllib.request.urlopen("http://127.0.0.1:5000")
html=resp.read().decode()
soup=BeautifulSoup(html,"lxml")
print("~")
elems=soup.select("div[class='quote']:nth-of-type(1) div[class='tags'] a:nth-of-type(1) ~ a")
for elem in elems:
    print(elem)
print("+")
elems=soup.select("div[class='quote']:nth-of-type(1) div[class='tags'] a:nth-of-type(1) + a")
for elem in elems:
    print(elem)
```

结果：

```
~
<a class="tag" href="/tag/deep-thoughts/page/1/">deep-thoughts</a>
<a class="tag" href="/tag/thinking/page/1/">thinking</a>
<a class="tag" href="/tag/world/page/1/">world</a>
+
<a class="tag" href="/tag/deep-thoughts/page/1/">deep-thoughts</a>
```

因为第一个<div class="quote">的<div class='tags'>中，第一个<a>后面的<a>节点有 3 个，所以使用 "～" 可以找到这 3 个节点，而使用 "+" 只能找到<a>后面的一个节点。

**例 2-5-7**：查找符合条件的兄弟节点。

```
import urllib.request
from bs4 import BeautifulSoup
import bs4
resp=urllib.request.urlopen("http://127.0.0.1:5000")
html=resp.read().decode()
soup=BeautifulSoup(html,"lxml")
elems=soup.select("div[class='quote']:nth-of-type(1) div meta ~ a[href*='world']")
for elem in elems:
    print(elem)
```

结果：

```
<a class="tag" href="/tag/world/page/1/">world</a>
```

因为<meta>后面 href 包含 "world" 的兄弟节点只有一个，因此所得的结果中只有一个节点。

### 2.5.6 使用 select_one() 查找单一元素

select_one(css)的查找规则与 select(css)的查找规则是一样的，只不过 select_one(css)只查找满足条件的第一个元素，如果找到就返回 bs4.element.Tag 对象，如果找不到就返回 None。

例 2-5-8：查找网站中第一个<a>元素。

```
import urllib.request
from bs4 import BeautifulSoup
resp=urllib.request.urlopen("http://127.0.0.1:5000")
html=resp.read().decode()
soup=BeautifulSoup(html,"lxml")
print(soup.select_one("a").text)
```

结果：

```
<a href="/author/Albert-Einstein">(about)</a>
```

其中，<a>元素很多，如果使用 soup.select("a")，那么可以找到所有的<a>节点。使用 soup.select_one("a")，只能找到第一个<a>节点。

## 2.6 综合项目 爬取模拟名言网站数据

### 任务目标

创建一个模拟名言网站，编写爬虫程序爬取其中的数据，并把数据存储在数据库中。通过本项目，我们能够学习使用 BeautifulSoup 解析与爬取网站的数据，为将来爬取真实网站的数据做准备。

### 2.6.1 创建模拟名言网站

创建网站模板文件 quotes.html，并创建网站服务器程序。运行服务器程序后访问"http://127.0.0.1:5000"，就会看到图 2-1-2 所示的模拟名言网站。

### 2.6.2 爬取名言数据

首先获取 HTML 文档，建立 BeautifulSoup 对象 soup。

```
resp=urllib.request.urlopen(url)
html=resp.read().decode()
soup=BeautifulSoup(html,"lxml")
```

#### 1. 爬取名言、名人文本

每条名言的 div 对象都是<div class="quote">元素，名言文本 quote 包含在第一个<span class="text">中，名人文本 author 包含在<small>中，代码如下：

```
divs=soup.select("div[class='quote']")
for div in divs:
    quote=div.select_one("span[class='text']").text
    author=div.select_one("small").text
```

#### 2. 爬取名言标签

每条名言都有多个标签 Tags，每个标签都是一个超链接，它们存在于<div class="tags">元素中，因此使用下面的方法获取。

```
divs=soup.select("div[class='quote']")
for div in divs:
    links= div.select("div[class='tags'] a")
    for link in links:
        print(link.text,link["href"])
```

### 2.6.3 设计存储数据库

设计一个 SQLite3 数据库 quotes.db 存储数据,这个数据库中有一张 quotes 表和一张 tags 表。因为一条名言往往有多个标签,所以单独使用 tags 表存储这些标签。quotes 表的各个字段如表 2-6-1 所示,tags 表的各个字段如表 2-6-2 所示。tags 表是 quotes 表的从表,tags 表通过 qID 字段与 quotes 表的 ID 关联。

表 2-6-1  quotes 表的各个字段

| 字段名称 | 类型 | 含义 |
| --- | --- | --- |
| ID | int | 自动增长列(关键字) |
| quote | varchar(1024) | 名言文本 |
| author | varchar(256) | 名人文本 |
| author_link | varchar(256) | 名人超链接 |

表 2-6-2  tags 表的各个字段

| 字段名称 | 类型 | 含义 |
| --- | --- | --- |
| ID | int | 自动增长列(关键字) |
| qID | int | quotes 表的 ID |
| tag | varchar(256) | 标签文本 |
| tag_link | varchar(256) | 标签超链接 |

### 2.6.4 编写爬虫程序

根据 HTML 代码分析,编写爬虫程序 spider.py,如下:

```python
import urllib.request
import bs4
from bs4 import BeautifulSoup
import sqlite3

class MySpider:

    def openDB(self):
        #初始化数据库
        self.con = sqlite3.connect("quotes.db")
        self.cursor = self.con.cursor()
        try:
            self.cursor.execute("drop table tags")
        except:
            pass
        try:
            self.cursor.execute("drop table quotes")
        except:
            pass
```

```python
            self.cursor.execute("create table quotes (ID INTEGER PRIMARY KEY AUTOINCREMENT ,quote varchar(1024),author varchar(256),author_link varchar(256))")
            self.cursor.execute("create table tags (ID INTEGER PRIMARY KEY AUTOINCREMENT ,qID int,tag varchar(256),tag_link varchar(256))")

    def insertQuote(self,quote,author,author_link):
        sql="insert into quotes (quote,author,author_link) values (?,?,?)"
        self.cursor.execute(sql,[quote,author,author_link])
        self.cursor.execute("select ID from quotes order by ID desc limit 1")
        row=self.cursor.fetchone()
        ID=row[0]
        return ID

    def insertTag(self,qID,tag,tag_link):
        sql="insert into tags (qID,tag,tag_link) values (?,?,?)"
        self.cursor.execute(sql,[qID,tag,tag_link])

    def closeDB(self):
        #关闭数据库
        self.con.commit()
        self.con.close()

    def spider(self,url):
        #爬虫函数
        try:
            resp=urllib.request.urlopen(url)
            html=resp.read().decode()
            soup=BeautifulSoup(html,"lxml")
            divs=soup.select("div[class='quote']")
            for div in divs:
                quote=div.select_one("span[class='text']").text
                author=div.select_one("small").text
                author_link=div.select_one("span:nth-of-type(2) a")["href"]
                ID=self.insertQuote(quote,author,author_link)
                links= div.select("div[class='tags'] a")
                for link in links:
                    self.insertTag(ID,link.text,link["href"])
        except Exception as err:
            print(err)

    def show(self):
        #显示函数
        self.cursor.execute("select ID,quote,author,author_link from quotes")
        quotes=self.cursor.fetchall()
        no=0
        for quote in quotes:
            no=no+1
            print("No",no)
            ID=quote[0]
            print(quote[1])
```

```
                print(quote[2])
                print(quote[3])
                self.cursor.execute("select qID,tag,tag_link from tags
where qID="+str(ID))
                tags=self.cursor.fetchall()
                for tag in tags:
                    print(tag[1]," --- ",tag[2])
                print()
            print("Total ",len(quotes))

#主程序
spider=MySpider()
spider.openDB()
spider.spider("http://127.0.0.1:5000")
spider.show()
spider.closeDB()
```

这个程序定义了一个爬虫类 MySpider,其中有几个主要的函数。

(1) openDB()函数:用于创建数据库连接对象 self.con 与游标对象 self.cursor,建立数据库连接;初始化 quotes 表与 tags 表,如果表已经存在则调用 drop table 命令删除,然后调用 create table 命令创建表,以保证每次爬取数据时该表是空的。

(2) insertQuote()函数:用于创建一条 quote 数据库记录,将数据存储到数据库,并返回这条记录的 ID。

(3) insertTag()函数:用于根据 quote 记录的 ID 插入一条 tags 表记录。

(4) closeDB()函数:在结束时调用该函数,用于关闭数据库。

(5) show()函数:用于显示数据库的 quotes 表与 tags 表中的数据。

(6) spider()函数:这是爬取数据的主要函数,它通过访问网页获取 HTML 代码,建立 BeautifulSoup 对象,爬取所需的数据,并将其存储到数据库。

### 2.6.5 执行爬虫程序

先执行服务器程序,再执行爬虫程序,成功爬取到如下数据。

```
No 1
"The world as we have created it is a process of our thinking. It cannot be changed without changing our thinking."
Albert Einstein
/author/Albert-Einstein
change  ---  /tag/change/page/1/
deep-thoughts  ---  /tag/deep-thoughts/page/1/
thinking  ---  /tag/thinking/page/1/
world  ---  /tag/world/page/1/

No 2
"It is our choices, Harry, that show what we truly are, far more than our abilities."
J.K. Rowling
/author/J-K-Rowling
abilities  ---  /tag/abilities/page/1/
choices  ---  /tag/choices/page/1/

Total  2
```

项目 ❷　爬取名言网站数据

## 2.7　实战项目　爬取实际名言网站数据

2-7-A　2-7-B
知识讲解　操作演练

**任务目标**

本项目的目标是学习使用 BeautifulSoup 爬取数据的方法，爬取网站中所有名人的名言，并将其存储到数据库。

### 2.7.1　解析网站的 HTML 代码

这里使用 Chrome 浏览器访问网站，在网站中找到一条名言，单击鼠标右键，在弹出的快捷菜单中选择"检查"命令，就可以看到图 2-7-1 所示的 HTML 代码。

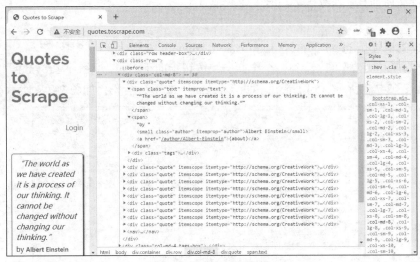

图 2-7-1　HTML 代码

我们可以看到名言的数据都在<div class="col-md-8">元素中，每条名言都在一个<div class="quote">元素中。复制一条名言的 HTML 代码，整理后得到：

```
        <div class="row">
            <div class="col-md-8">
            <div class="quote" itemscope="" itemtype="http://schema.org/CreativeWork">
                <span class="text" itemprop="text">"The world as we have created it is a process of our thinking. It cannot be changed without changing our thinking."</span>
                <span>by <small class="author" itemprop="author">Albert Einstein</small>
                <a href="/author/Albert-Einstein">(about)</a>
                </span>
                <div class="tags">
                    Tags:
                    <meta class="keywords" itemprop="keywords" content="change,deep-thoughts,thinking,world">
                        <a class="tag" href="/tag/change/page/1/">change</a>
                        <a class="tag" href="/tag/deep-thoughts/page/
```

55

```
1/">deep-thoughts</a>
                <a class="tag" href="/tag/thinking/page/1/">thinking</a>
                <a class="tag" href="/tag/world/page/1/">world</a>
            </div>
        </div>
        ……
    </div>
</div>
```

这个结构与模拟名言网站的结构类似,因此爬取数据的方法非常类似,这里不赘述。

### 2.7.2 爬取全部页面的数据

查看网页时发现:

第 1 页的地址是 http://quotes.toscrape.com/page/1/;

第 2 页的地址是 http://quotes.toscrape.com/page/2/;

……

第 $n$ 页的地址是 http://quotes.toscrape.com/page/n/。

而且在页面的底部可以看到一个"Next"按钮指向下一页,一个"Previous"按钮指向前一页,如图 2-7-2 所示。到了最后一页,"Next"按钮会消失,如图 2-7-3 所示。

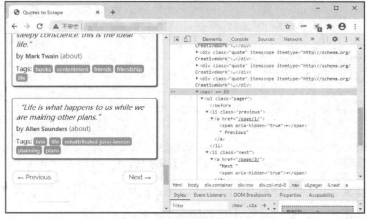

图 2-7-2　页面翻页

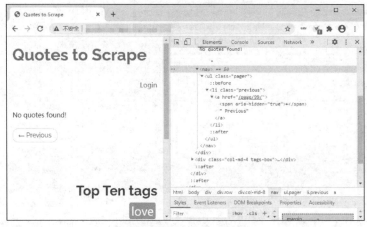

图 2-7-3　最后一页

## 项目 ❷ 爬取名言网站数据

根据这个特征,可以设计一个 $n$ 变量页面计算器,$n$ 从 1 开始递增,依次爬取页面 http://quotes.toscrape.com/page/n/ 的数据,当 $n$ 递增到某个页面找不到 "Next" 按钮时,说明到了最后一页,爬取过程就停止了。

根据网站 HTML 代码的特征,显然可以使用下面的语句查找 "Next" 按钮。

```
soup.select("nav ul[class='pager'] li[class='next']")
```

### 2.7.3 编写爬虫程序

根据 HTML 代码分析,编写爬虫程序 spider.py,如下:

```python
import urllib.request
import bs4
from bs4 import BeautifulSoup
import sqlite3

class MySpider:

    def initDB(self):
        #初始化数据库
        self.con = sqlite3.connect("quotes.db")
        self.cursor = self.con.cursor()
        try:
            self.cursor.execute("drop table tags")
        except:
            pass
        try:
            self.cursor.execute("drop table quotes")
        except:
            pass
        self.cursor.execute("create table quotes (ID INTEGER PRIMARY KEY AUTOINCREMENT ,quote varchar(1024),author varchar(256),author_link varchar(256))")
        self.cursor.execute("create table tags (ID INTEGER PRIMARY KEY AUTOINCREMENT ,qID int,tag varchar(256),tag_link varchar(256))")

    def openDB(self):
        #打开数据库
        self.con = sqlite3.connect("quotes.db")
        self.cursor = self.con.cursor()

    def insertQuote(self,quote,author,author_link):
        sql="insert into quotes (quote,author,author_link) values (?,?,?)"
        self.cursor.execute(sql,[quote,author,author_link])
        self.cursor.execute("select ID from quotes order by ID desc limit 1")
        row=self.cursor.fetchone()
        ID=row[0]
        return ID

    def insertTag(self,qID,tag,tag_link):
        sql="insert into tags (qID,tag,tag_link) values (?,?,?)"
        self.cursor.execute(sql,[qID,tag,tag_link])
```

```python
    def closeDB(self):
        #关闭数据库
        self.con.commit()
        self.con.close()

    def spider(self,url):
        #爬虫函数
        print(url)
        try:
            resp=urllib.request.urlopen(url)
            html=resp.read().decode()
            soup=BeautifulSoup(html,"lxml")
            divs=soup.select("div[class='col-md-8'] div[class='quote']")
            for div in divs:
                quote=div.select_one("span[class='text']").text
                author=div.select_one("small").text
                author_link=div.select_one("span:nth-of-type(2) a")["href"]
                ID=self.insertQuote(quote,author,author_link)
                links= div.select("div[class='tags'] a")
                for link in links:
                    self.insertTag(ID,link.text,link["href"])
            return soup.select("nav ul[class='pager'] li[class='next']")
        except Exception as err:
            print(err)

    def process(self):
        self.initDB()
        n=0
        con=True
        while con:
            n=n+1
            url="http://quotes.toscrape.com/page/"+str(n)+"/"
            con=self.spider(url)
        self.closeDB()

    def show(self):
        #显示函数
        self.openDB()
        self.cursor.execute("select ID,quote,author,author_link from quotes")
        quotes=self.cursor.fetchall()
        no=0
        for quote in quotes:
            no=no+1
            print("No",no)
            ID=quote[0]
            print(quote[1])
            print(quote[2])
            print(quote[3])
            self.cursor.execute("select qID,tag,tag_link from tags
```

```
where qID="+str(ID))
                tags=self.cursor.fetchall()
                for tag in tags:
                    print(tag[1]," --- ",tag[2])
                print()
        print("Total ",len(quotes))
        self.closeDB()

#主程序
spider=MySpider()
while True:
    print("1.爬取")
    print("2.显示")
    print("3.退出")
    s=input("选择(1,2,3):")
    if s=="1":
        spider.process()
    elif s=="2":
        spider.show()
    elif s=="3":
        break
```

这个程序定义了一个爬虫类 MySpider，其中有几个主要的函数。

（1）initDB()函数：用于初始化 quotes 表与 tags 表，如果表已经存在，则调用 drop table 命令删除，然后调用 create table 命令创建表，以保证每次爬取数据时该表是空的。

（2）openDB()函数：用于创建数据库连接对象 self.con 与游标对象 self.cursor，建立数据库连接。

（3）closeDB()函数：用于关闭数据库。

（4）spider()函数：这是程序的核心函数，用于爬取网页的数据，并在最后判断是否能找到"Next"按钮。

（5）process()函数：用于循环爬取各个网页的数据，如果找不到"Next"按钮就停止。

（6）show()函数：用于显示数据库中的记录。

### 2.7.4 执行爬虫程序

（1）执行爬虫程序，选择"1"爬取数据，可以看到 10 个页面的地址。

```
http://quotes.toscrape.com/page/1/
http://quotes.toscrape.com/page/2/
http://quotes.toscrape.com/page/3/
http://quotes.toscrape.com/page/4/
http://quotes.toscrape.com/page/5/
http://quotes.toscrape.com/page/6/
http://quotes.toscrape.com/page/7/
http://quotes.toscrape.com/page/8/
http://quotes.toscrape.com/page/9/
http://quotes.toscrape.com/page/10/
```

（2）选择"2"显示这些数据，发现有100条名言记录，下面是部分结果。

```
No 1
"The world as we have created it is a process of our thinking. It cannot
be changed without changing our thinking."
Albert Einstein
/author/Albert-Einstein
change       ---   /tag/change/page/1/
deep-thoughts  ---   /tag/deep-thoughts/page/1/
thinking     ---   /tag/thinking/page/1/
world        ---   /tag/world/page/1/
……
No 100
"... a mind needs books as a sword needs a whetstone, if it is to keep its edge."
George R.R. Martin
/author/George-R-R-Martin
books        ---   /tag/books/page/1/
mind         ---   /tag/mind/page/1/

Total  100
```

值得注意的是，该网站的各个页面的地址正好是有规律的，因此可以很容易地构造各个页面的地址，然后爬取各个页面的数据。如果各个页面的地址没有规律，就不能这样简单地循环爬取了，但可以使用递归方法来爬取，在后面的项目中会介绍这种递归方法。

## 项目总结

本项目涉及一个名言网站，我们使用 BeautifulSoup 对网站的网页进行解析并得到所要的数据，实现了爬取网站数据的爬虫程序。

BeautifulSoup 是一个功能强大的网页解析工具，一般使用它的 find() 函数与 find_all() 函数就可以找到所要的数据，它还支持 CSS 语法的查找方法，因此使用它查找数据十分方便。

实际上，一个网站有很多网页，数据往往分布在不同的网页中，爬虫程序应该能自如地遍历各个网页并爬取数据，在后面的项目中将讲解高效地爬取多个网页数据的爬虫技术。

## 练习 2

1. 简单说明 BeautifulSoup 解析数据的特点。
2. 用 BeautifulSoup 装载下面的 HTML 文档，并以规范的格式输出，比较与原来 HTML 文档的区别，说明 BeautifulSoup 是如何修改的。

```
<body>
<div>Hi<br>
<span>Hello</SPAN>
<p><div>End</div>
```

3. 重新编写本书项目 1 中爬取外汇网站数据的程序，使用 BeautifulSoup 分解出<tr>…</tr>中的<td>…</td>数据。

4. 下面是一段 HTML 代码：

```
<body>
```

```
<bookstore>
<book id="b1">
  <title lang="english">Harry Potter</title>
  <price>23.99</price>
</book>
<book id="b2">
  <title lang="chinese">学习 XML</title>
  <price>39.95</price>
</book>
<book id="b3">
  <title lang="english">Learning Python</title>
  <price>30.20</price>
</book>
</bookstore>
</body></html>
```

试用 BeautifulSoup 完成下面的任务。

（1）找出所有书的名称。

（2）找出所有英文书的名称与价格。

（3）找出价格在 30 元以上的所有书的名称。

5．编写一个爬取 Python 最新版本的 Windows 64 位压缩包的程序，程序分两个部分。

（1）访问 Python 下载页面 "https://www.python.org/downloads/"，如图 2-9-1 所示。

图 2-9-1　Python 下载页面

分析 HTML 代码结构，编写下面的程序，爬取所有发行的 Python 版本与下载地址。

```
from bs4 import  BeautifulSoup
import urllib.request
def searchPython(url):
    resp=urllib.request.urlopen(url)
    data=resp.read()
    html=data.decode()
    soup=BeautifulSoup(html,"lxml")
    ol=soup.find(name="ol",attrs={"class":"list-row-container menu"})
    lis=ol.find_all("li")
    for li in lis:
        a=li.find(name="span",attrs={"class":"release-number"}).find("a")
```

```
            python=a.text
            url=a["href"]
            print("%-20s %s" %(python,url))
try:
    searchPython("https://www.python.org/downloads/")
except Exception as e:
    print(e)
```

执行该程序，部分结果如下：

```
Python 2.5.4            /download/releases/2.5.4/
Python 2.4.6            /download/releases/2.4.6/
Python 2.5.3            /download/releases/2.5.3/
Python 2.6.1            /download/releases/2.6.1/
Python 3.0.0            /download/releases/3.0/
......
```

（2）在这个程序的基础上进一步编写程序，找出最新的版本，自动进入最新版本下载页面，找出这个版本的 Windows 64 位的 ZIP 压缩包的下载地址，并自动下载这个 Python 压缩包。

6. 在中国天气网中查找一个城市，如深圳，会转到地址为"http://www.weather.com.cn/weather/101280601.shtml"的网页显示深圳的天气预报，如图 2-9-2 所示。

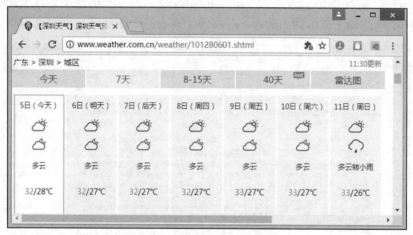

图 2-9-2　深圳的天气预报

编写爬虫程序爬取该网站 7 天内的天气预报，包括日期、天气、温度等数据。

# 项目 ③ 爬取电影网站数据

复杂的爬虫程序爬取的数据往往很多，而且相关的数据往往分布在很多不同的网页中，爬虫程序必须能按链接自动往返于这些不同的网站去爬取数据。一个爬虫程序爬取成百上千条的数据是常有的事，因此如何设计一个高效率的爬虫程序成了本项目学习的重点。

本项目中我们将爬取豆瓣电影网站经典的 250 部电影数据，这些电影中有很多优秀的国产电影。改革开放以来，国产影片的创作类型不断拓展，国产电影透过镜头，以更多元化、更有影响力的形态，向世界彰显中国的文化自信。

拓展阅读

网络爬虫与法律法规

## 3.1 项目任务

豆瓣电影网站列出了经典的 250 部电影的简介，进入该网站，找到"排行榜"下的"豆瓣电影 Top 250"，单击"全部"即可看到这些电影的信息，如图 3-1-1 所示。由于电影比较多，一个页面只列出 25 部电影的简介，全部电影分为 10 个页面，单击网页下面的"后页>"链接就可以跳转到下一个页面。设计一个爬虫程序，它在爬取一个页面后能自动跳转到下一个页面继续爬取，程序结束时能爬取所有 250 部电影的信息。

在爬取实际网站的数据之前先练习爬取模拟网站的数据。创建一个项目文件夹 project3，在该文件夹中有一个 movies.csv 文件，其中包含很多电影的信息，前面几行如下：

3-1-A  3-1-B

知识讲解　操作演练

图 3-1-1　豆瓣电影网站

```
ID,mImage,mTitle,mNative,mNickname,mDirectors,mActors,mTime,mCountry,mType,mPoint,mComments
000001,.jpg,肖申克的救赎,The Shawshank Redemption,月黑高飞 / 刺激 1995,弗兰克·德拉邦特 Frank Darabont,蒂姆·罗宾斯 Tim Robbins /...,1994，美国，犯罪 剧情,9.6,1241123 人
000002,.jpg,霸王别姬,Farewell My Concubine,陈凯歌 Kaige Chen ,张国荣 Leslie Cheung / 张丰毅 Fengyi Zha...,1993，中国，剧情 爱情,9.6,914326 人
……
```

文件中的各个数据字段使用","分隔，第一行是各个数据字段的名称，其中 ID 是编号、mImage 是电影的图像扩展名、mTitle 是名称、mNative 是原名、mNickname 是别名、

mDirectors 是导演、mActors 是主演、mTime 是时间（指上映时间）、mCountry 是国家、mType 是类型、mPoint 是评分、mComments 是评价（指评价人数）。

在 images 文件夹中有各个电影的图像，图像的名称是以电影编号命名的，如图 3-1-2 所示。

图 3-1-2　电影图像

我们使用这些数据编写一个模拟电影网站，如图 3-1-3 所示。这个网站可以显示 250 部经典电影的信息，信息分布在很多页面，单击每个网页的"第一页""前一页""下一页" "末一页"等按钮就可以在各个页面之间切换。

图 3-1-3　模拟电影网站

## 3.2　简单爬取网站数据

知识讲解

操作演练

### 任务目标

使用 movies.csv 文件中的数据，根据真实的电影网站的布局创建一个模拟电影网站，分析网站的 HTML 代码并使用 BeautifulSoup 爬取电影数据。在本节中，我们主要学习设计爬虫程序以爬取图像等复杂的数据。

## 3.2.1 创建模拟电影网站

### 1. 创建网站模板

在 project3 中创建一个名称为 templates 的子文件夹，再依据 movies.csv 文件中的一行数据在该子文件夹中创建一个 movie.html 文件，该文件的内容如下：

```html
<style>
.pic {display:inline-block;width:130px; vertical-align:top;margin:10px;}
.info { display:inline-block; }
.liClass { list-style: none; margin:20px;}
.title { margin: 10px; font-size:18px;}
.point { margin: 10px;color:red; }
.attrs { margin: 10px;color: #666; }
h3 { display:inline-block;}}
.pl {color:#888;}
.native { fons-size:10px;}
.nickname { fons-size:10px;}
.directors { fons-size:10px;}
.actors { fons-size:10px;}
,others { fons-size:10px; }
</style>
<body>
  <div class="pic">
     <img width="100"  src="images/000001.jpg">
  </div>
  <div class="info">
        <div class="title"><h3 style="display:inline-block">肖申克的救赎</h3>  <span class="point">9.6分</span></div>
        <div class="native">
             <span class="pl">原名</span>:<span class="attrs">The Shawshank Redemption</span>
        </div>
        <div class="nickname">
             <span class="pl">别名</span>:<span   class="attrs">月黑高飞  / 刺激1995</span>
        </div>
        <div class="directors">
             <span class="pl">导演</span>:<span class="attrs">弗兰克·德拉邦特 Frank Darabont</span>
        </div>
        <div class="actors">
             <span class="pl">主演</span>:<span class="attrs">蒂姆·罗宾斯 Tim Robbins/...</span>
        </div>
         <div class="others">
             <span class="pl">时间</span>:<span class="attrs"> 1994</span><br>
             <span class="pl">国家</span>:<span class="attrs"> 美国</span><br>
             <span class="pl">类型</span>:<span class="attrs"> 犯罪 剧情</span><br>
```

```
                <span class="pl">评价</span>:<span class="attrs">1241123 人
</span>
            </div>
        </div>
```

### 2. 创建网站服务器程序

在 project3 中编写服务器程序 server.py。

```python
import flask
app=flask.Flask(__name__,static_folder="images")
@app.route("/")
def index():
    return flask.render_template("movie.html")
app.run()
```

其中，使用 static_folder="images"指定了网站的静态文件夹是 images，这是为了显示图像。因为客户端访问图像时使用的是地址 "http://127.0.0.1:5000/images/000001.jpg"，所以 Flask 服务器必须知道 images 是一个静态文件夹，只有这样，Flask 服务器才会向客户端提交 000001.jpg 图像。运行该服务器程序，使用浏览器访问 "http://127.0.0.1:5000"，结果如图 3-2-1 所示。

图 3-2-1 模拟电影网站

## 3.2.2 爬取网站数据

### 1. 获取 HTML 代码

编写客户端程序 client.py，如下：

```python
import urllib.request
from bs4 import BeautifulSoup
try:
    resp=urllib.request.urlopen("http://127.0.0.1:5000")
    html=resp.read().decode()
    soup=BeautifulSoup(html,"lxml")
    #解析 HTML 代码
except Exception as err:
    print(err)
```

这个程序可以获取 HTML 代码并建立 BeautifulSoup 对象 soup。

### 2. 爬取电影名称

分析网站的 HTML 代码可以看到，所有的电影数据都包含在<div class="info">中，而电影名称包含在<div class="title">的<h3>中，原名包含在<div class="native">的<span class=

"attrs">中,别名包含在<div class="nickname">的<span class="attrs">中,因此通过下列程序爬取电影名称。

```
div=soup.find("div",attrs={"class":"info"})
mTitle=div.find("div",attrs={"class":"title"}).find("h3").text
mNative = div.find("div", attrs={"class": "native"}).find
("span",attrs={"class":"attrs"}).text
mNickname = div.find("div", attrs={"class": "nickname"}).find("span",
attrs={"class":"attrs"}).text
print(mTitle,",",mNative,",",mNickname)
```

### 3. 爬取导演与主演名字

导演名字包含在<div class="dorectors">的<span class="attrs">中,主演名字包含在<div class="actors">的<span class="attrs">中,因此通过下列程序爬取导演与主演名字。

```
div=soup.find("div",attrs={"class":"info"})
mDirectors = div.find("div", attrs={"class": "directors"}).find("span",
attrs={"class":"attrs"}).text
mActors = div.find("div", attrs={"class": "actors"}).find("span",
attrs={"class":"attrs"}).text
print(mDirectors,",",mActors)
```

### 4. 爬取其他信息

其他信息包含在<div class="others">中,时间、国家、类型、评价依次包含在各个<span class="attrs">中,因此通过下列程序爬取其他信息。

```
spans=div.find("div", attrs={"class": "others"}).find_all("span",
attrs={"class":"attrs"})
mTime=spans[0].text
mCountry=spans[1].text
mType=spans[2].text
mComments=spans[3].text
```

### 5. 爬取电影图像

电影图像包含在<div class="pic"> <img width="100"  src="images/000001.jpg"></div>中。首先得到<img>的 src 值"images/000001.jpg",这是图像的地址,不过这是相对地址,还需要使用 urllib.request.urljoin(ur,src)把它转换为绝对地址,即:

```
src=soup.find("div",attrs={"class":"pic"}).find("img")["src"]
src=urllib.request.urljoin(url,src)
```

编写一个下载图像的函数,如下:

```
def download(src):
    try:
        p=src.rfind("/")
        name=src[p+1:]
        resp=urllib.request.urlopen(src)
        data=resp.read()
        f=open("download\\"+name,"wb")
        f.write(data)
        f.close()
        print("download ",name)
    except Exception as err:
        print(err)
```

其中，src 是要下载的图像地址，函数找到地址的最后一个"/"，"/"后面的部分就是图像的名称 name。使用 urlopen(src)打开网页，读取的二进制数据就是图像的数据 data，把数据存储到 download 文件夹下就完成了图像下载。

### 3.2.3 编写爬虫程序

根据前面的分析，编写爬虫程序 spider.py，如下：

```python
import urllib.request
from bs4 import BeautifulSoup
import os

def spider(url):
    m = {}
    try:
        resp=urllib.request.urlopen(url)
        html=resp.read().decode()
        soup=BeautifulSoup(html,"lxml")
        #解析HTML代码
        div=soup.find("div",attrs={"class":"info"})
        m["mTitle"]=div.find("div",attrs={"class":"title"}).find("h3").text
        m["mNative"] = div.find("div", attrs={"class": "native"}).find("span",attrs={"class":"attrs"}).text
        m["mNickname"] = div.find("div", attrs={"class": "nickname"}).find("span",attrs={"class":"attrs"}).text
        m["mDirectors"] = div.find("div", attrs={"class": "directors"}).find("span",attrs={"class":"attrs"}).text
        m["mActors"] = div.find("div", attrs={"class": "actors"}).find("span",attrs={"class":"attrs"}).text
        spans=div.find("div", attrs={"class": "others"}).find_all("span",attrs={"class":"attrs"})
        m["mTime"]=spans[0].text
        m["mCountry"]=spans[1].text
        m["mType"]=spans[2].text
        m["mComments"]=spans[3].text
        src=soup.find("div",attrs={"class":"pic"}).find("img")["src"]
        src=urllib.request.urljoin(url,src)
        download(src)
    except Exception as err:
        print(err)
    return m

def download(src):
    try:
        p=src.rfind("/")
        name=src[p+1:]
        resp=urllib.request.urlopen(src)
        data=resp.read()
        f=open("download\\"+name,"wb")
        f.write(data)
        f.close()
        print("download ",name)
    except Exception as err:
```

```
            print(err)
def show(m):
    print("名称: ",m["mTitle"])
    print("原名: ", m["mNative"])
    print("别名: ", m["mNickname"])
    print("导演: ",m["mDirectors"])
    print("主演: ",m["mActors"])
    print("时间: ",m["mTime"])
    print("国家: ",m["mCountry"])
    print("类型: ",m["mType"])
    print("评价: ",m["mComments"])

if not os.path.exists("download"):
    os.mkdir("download")
m=spider("http://127.0.0.1:5000")
show(m)
```

### 3.2.4 执行爬虫程序

执行爬虫程序后，可以看到成功爬取到该电影的各个数据，还爬取到电影的图像000001.jpg，并将其存储在download子文件夹下，结果如下：

```
download 000001.jpg
名称：   肖申克的救赎
原名：   The Shawshank Redemption
别名：   月黑高飞  /  刺激1995
导演：   弗兰克·德拉邦特 Frank Darabont
主演：   蒂姆·罗宾斯 Tim Robbins/...
时间：   1994
国家：   美国
类型：   犯罪 剧情
评价：   1241123 人
```

## 3.3 递归爬取网站数据

3-3-A
知识讲解

3-3-B
操作演练

**任务目标**

首先，使用已有数据根据真实电影网站创建一个模拟电影网站，这个网站包含多个网页，网页之间通过超链接相互关联；然后，编写一个爬虫程序，自动遍历每个网页，爬取每个网页的数据。在本节中，我们主要学习使用递归程序遍历每个网页的方法。

### 3.3.1 创建模拟电影网站

**1. 创建网站模板**

使用movies.csv文件中的数据并仿照movie.html文件的格式，使用电影名称分别创建

下面的网页文件:
  (1)电影肖申克的救赎.html;
  (2)霸王别姬.html;
  (3)这个杀手不太冷.html;
  (4)阿甘正传.html;
  (5)千与千寻.html。
  然后创建下面几个网页文件。
  (1)movie.html:

```html
<body>
<div id="country">电影网站</div>
<ul>
    <li><a href="美国电影.html">美国电影</a></li>
    <li><a href="中国电影.html">中国电影</a></li>
    <li><a href="法国电影.html">法国电影</a></li>
    <li><a href="日本电影.html">日本电影</a></li>
</ul>
</body>
```

  (2)美国电影.html:

```html
<body>
<div id="country">美国电影</div>
<ul>
    <li><a href="肖申克的救赎.html">肖申克的救赎</a></li>
    <li><a href="阿甘正传.html">阿甘正传 </a></li>
    <div><a href="/">【返回】</a></div>
</ul>
</body>
```

  (3)中国电影.html:

```html
<body>
<div id="country">中国电影</div>
<ul>
    <li><a href="霸王别姬.html">霸王别姬</a></li>
    <div><a href="/">【返回】</a></div>
</ul>
</body>
```

  (4)法国电影.html:

```html
<body>
<div id="country">法国电影</div>
<ul>
    <li><a href="这个杀手不太冷.html">这个杀手不太冷</a></li>
        <div><a href="/">【返回】</a></div>
</ul>
</body>
```

  (5)日本电影.html:

```html
<body>
```

```
<div id="country">日本电影</div>
<ul>
    <li><a href="千与千寻.html">千与千寻</a></li>
    <div><a href="/">【返回】</a></div>
</ul>
</body>
```

## 2. 创建网站服务器程序

在 project3 中编写服务器程序 server.py，如下：

```
import flask
app=flask.Flask(__name__,static_folder="images")

@app.route("/")
def index():
    return flask.render_template("movie.html")

@app.route("/<name>")
def show(name):
    if name.strip()=="":
        name="movie.html"
    return flask.render_template(name)

app.run()
```

这个服务器设置"http://127.0.0.1:5000"的入口网页是 movie.html，而显示某个网页时，例如，显示"http://127.0.0.1:5000/中国电影.html"时，这个网址后面的部分"中国电影.html"就被当成参数 name 传递给 show()函数，并在 show()函数中提交这个网页。

运行服务器程序，然后使用浏览器访问"http://127.0.0.1:5000"，就会看到图 3-3-1 所示的页面，单击任何一个超链接就进入相应的具体电影名称页面，如图 3-3-2 所示。

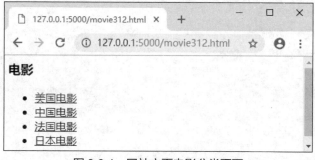

图 3-3-1　网站主页电影分类页面

图 3-3-2　具体电影名称页面

### 3.3.2 解析电影网站结构

这个网站虽然小，但是包含多个网页，网站的结构如图 3-3-3 所示。这个结构实际上是树状的图形结构，顶层是电影网站，第二层是电影分类，第三层是具体的电影名称。

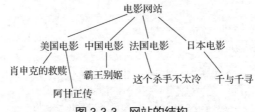

图 3-3-3 网站的结构

实际上，任何一个网站的网页与网页之间，甚至网站与网站之间，都是通过超链接相互关联的，组成一个复杂的有向图。网站的每个网页组成图的节点，超链接是图的边，数据就分布在这张图的各个网页节点上。

一个爬虫程序要爬取数据，经常要在这张图中从一个网页节点到另外一个网页节点，而如何遍历这张图的每个网页节点且不重复是我们关心的问题。

### 3.3.3 递归爬取电影网站数据

爬虫程序爬取网站中各个网页的数据时可以采用递归程序的方法实现，具体步骤如下：

（1）设计一个全局的列表变量 visited=[]；

（2）访问一个网页，使用 visited.append(url)记录该网页已经访问过，下次不再访问；

（3）爬取这个网页的数据；

（4）获取这个网页中所有<a>的 href 值，并形成新的地址，如果地址在 visited 中出现过就不再进行，否则回到（2）。

这个方法实际上是有向图的节点递归遍历方法，根据这个方法编写递归的爬虫程序 spider.py，如下：

```
import urllib.request
from bs4 import BeautifulSoup

def spider(url):
    global visited
    try:
        #print(url)
        visited.append(url)
        resp = urllib.request.urlopen(url)
        html = resp.read().decode()
        soup = BeautifulSoup(html, "lxml")
        div = soup.find("div", attrs={"class": "info"})
        if div:
            mTitle = div.find("div", attrs={"class": "title"}).find("h3").text
            print("---",mTitle)
        else:
            div = soup.find("div", attrs={"id": "country"})
            print(div.text)
```

```
            links = soup.find_all("a")
            for link in links:
                    href=urllib.parse.quote(link["href"])
                    url=urllib.request.urljoin(url,href)
                    if not url in visited:
                            spider(url)
    except Exception as err:
            print(err)

visited=[]
spider("http://127.0.0.1:5000/")
```

简单起见，程序在爬取具体的电影网站时只爬取它的名称，在爬取电影网页时只爬取它的<div id="country">的文本值。程序执行后结果如下：

```
电影网站
美国电影
--- 肖申克的救赎
--- 阿甘正传
中国电影
--- 霸王别姬
法国电影
--- 这个杀手不太冷
日本电影
--- 千与千寻
```

这个结果表明，程序遍历了每个网页，而且爬取到了每个网页的数据。

## 3.4 深度优先爬取网站数据

3-4-A 知识讲解　　3-4-B 操作演练

 任务目标

使用递归的方法可以遍历各个相关的网页并爬取其中的数据。实际上，递归的本质就是深度优先法，它与数据结构中的栈密切相关。在本节中，我们主要学习栈的使用方法与深度优先法，通过深度优先法来遍历各个网页并爬取其中的数据。

### 3.4.1 深度优先法

要使用深度优先法爬取网站数据，可以设计一个栈 Stack。在 Python 中实现一个栈十分简单，Python 中的列表就是一个栈。设计一个 Stack 类，如下：

```
class Stack:
    def __init__(self):
        self.st=[]
    def pop(self):
        return self.st.pop()
    def push(self,obj):
        self.st.append(obj)
    def empty(self):
        return len(self.st)==0
```

其中,push()是入栈函数,pop()是出栈函数,empty()函数用于判断栈是否为空。根据深度优先规则爬取数据的爬虫程序的运行过程如下:

(1)设计一个全局的列表变量 visited=[];

(2)第一个地址入栈;

(3)如果栈为空则程序结束,如果栈不为空则让一个地址出栈,爬取其中的数据;

(4)获取 url 站点的所有<a>的 href 值并组成新的地址,只要这个地址没有在 visited 中出现过就将其入栈;

(5)回到(3)。

### 3.4.2　深度优先爬虫程序

仍然采用 3.3 节所示的网站,根据深度优先规则编写爬虫程序 spider.py,如下:

```python
import urllib.request
from bs4 import BeautifulSoup

class Stack:
    def __init__(self):
        self.st=[]
    def pop(self):
        return self.st.pop()
    def push(self,obj):
        self.st.append(obj)
    def empty(self):
        return len(self.st)==0

def spider(url):
    try:
        visited = []
        stack=Stack()
        stack.push(url)
        visited.append(url)
        while not stack.empty():
            url=stack.pop()
            resp = urllib.request.urlopen(url)
            html = resp.read().decode()
            soup = BeautifulSoup(html, "lxml")
            div = soup.find("div", attrs={"class": "info"})
            if div:
                mTitle = div.find("div", attrs={"class": "title"}).find("h3").text
                print("---",mTitle)
            else:
                div = soup.find("div", attrs={"id": "country"})
                print(div.text)
            links = soup.find_all("a")
            for link in links:
                href=urllib.parse.quote(link["href"])
                url=urllib.request.urljoin(url,href)
                if not url in visited:
                    stack.push(url)
```

```
                visited.append(url)
        except Exception as err:
            print(err)

spider("http://127.0.0.1:5000/")
```
执行该程序，结果如下。

```
电影网站
日本电影
--- 千与千寻
法国电影
--- 这个杀手不太冷
中国电影
--- 霸王别姬
美国电影
--- 阿甘正传
--- 肖申克的救赎
```

由此可见，结果与使用递归方法得到的结果差不多，只是顺序上有些不同。这个顺序主要是由入栈的过程确定的。如果把入栈的顺序程序：

```
links = soup.find_all("a")
for link in links:
    href=urllib.parse.quote(link["href"])
```

修改为：

```
links = soup.find_all("a")
for i in range(len(links)-1,-1,-1):
    link=links[i]
    href=urllib.parse.quote(link["href"])
```

再次执行程序，得到的结果就与使用递归方法得到的结果完全一样了。为什么会这样呢？实际上，递归的本质就是深度优先法，程序的递归是通过栈实现的。

## 3.5 广度优先爬取网站数据

3-5-A 知识讲解　　3-5-B 操作演练

**任务目标**

使用深度优先法爬取网页是向网页树形结构的纵深方向进行的，而在一个实际的网站中，往往前面几级关联的网页才是比较重要的，因此使用深度优先法有一定的局限性。如果重点关注前面几级关联网页的数据，使用广度优先法会更加合适。广度优先法与数据结构中的队列密切相关。在本节中，我们主要学习使用队列与广度优先法遍历各个网页的方法。

### 3.5.1 广度优先法

遍历网站时还有一种广度优先的顺序，需要用到队列。在 Python 中实现一个队列十分简单，Python 中的列表就是一个队列。因此，很容易设计一个 Queue 类：

```
class Queue:
```

```
        def __init__(self):
            self.st=[]
        def fetch(self):
            return self.st.pop(0)
        def enter(self,obj):
            self.st.append(obj)
        def empty(self):
            return len(self.st)==0
```

其中，enter()是入队函数，fetch()是出队函数，empty()用于判断队列是否为空。采用 Queue 类后，按照广度优先的顺序爬取数据。客户端程序的运行过程如下：

（1）设计一个全局的列表变量 visited=[]；
（2）第一个地址入队；
（3）如果队列为空则程序结束，如果队列不为空则让一个地址出队，爬取其中的数据；
（4）获取 url 站点的所有<a>的 href 值并组成新的地址，如果地址没有在 visited 中出现过就将其入队，否则抛弃；
（5）回到（3）。

### 3.5.2 广度优先爬虫程序

仍然采用 3.3 节所示的网站，根据广度优先规则编写爬虫程序 spider.py，如下：

```
import urllib.request
from bs4 import BeautifulSoup

class Queue:
    def __init__(self):
        self.st=[]
    def fetch(self):
        return self.st.pop(0)
    def enter(self,obj):
        self.st.append(obj)
    def empty(self):
        return len(self.st)==0

def spider(url):
    try:
        visited = []
        queue=Queue()
        queue.enter(url)
        visited.append(url)
        while not queue.empty():
            url=queue.fetch()
            resp = urllib.request.urlopen(url)
            html = resp.read().decode()
            soup = BeautifulSoup(html, "lxml")
            div = soup.find("div", attrs={"class": "info"})
            if div:
                mTitle = div.find("div", attrs={"class": "title"}).find("h3").text
                print("---",mTitle)
            else:
```

```
                div = soup.find("div", attrs={"id": "country"})
                print(div.text)
            links = soup.find_all("a")
            for link in links:
                href=urllib.parse.quote(link["href"])
                url=urllib.request.urljoin(url,href)
                if not url in visited:
                    queue.enter(url)
                    visited.append(url)
    except Exception as err:
        print(err)

spider("http://127.0.0.1:5000/")
```

执行该程序，结果如下：

电影网站
美国电影
中国电影
法国电影
日本电影
--- 肖申克的救赎
--- 阿甘正传
--- 霸王别姬
--- 这个杀手不太冷
--- 千与千寻

使用广度优先法得到的结果与使用深度优先法得到的结果顺序不同。广度优先法是按图的生成树分层遍历的，这种方法的一个好处是可以只访问主页下面几层的节点，而且很容易控制访问的节点层数，不会使程序爬到很深层的网页去，这对于爬取很复杂的网站是很重要的。

## 3.6 爬取翻页网站数据

3-6-A  3-6-B

知识讲解  操作演练

### 任务目标

在很多网站中，同一个类型的栏目往往被组织在很多个网页中，网页可以通过"第一页""前一页""下一页""末一页"等按钮在各个页面中跳转，如图 3-6-1 所示。在本节中，我们主要学习爬取这种类型的网站中各个网页的数据的方法。

### 3.6.1 使用 Flask 模板参数

为了创建一个有翻页按钮的网站，必须先了解 Flask 中的参数传递问题。在 Flask 的 Web 服务器中，templates 是模板文件夹，该文件夹中的 HTML 文件是可以通过 flask.render_template()函数返回给客户端的。

图 3-6-1 有翻页按钮的网页

现在有几百部电影，是不是要为每部电影都设计一个网页呢？实际上没有必要，Flask 允许使用模板文件，文件中含有一些参数，把一部电影的参数传递给这个模板文件就可以得到该电影的网页，如果将另外一部电影的参数传递给这个模板文件，又可以得到另外一部电影的网页，因此只要设计好这个模板文件，为它传递不同的参数就可以得到不同电影的网页了。

**1. 模板使用字符串参数**

向模板文件传递字符串参数是常用的。现在设计一个网页模板文件 movie.html，该文件内容如下：

```html
<body>
    名称:{{mTitle}}<br>
    导演:{{mDirectors}}<br>
    主演:{{mActors}}
</body>
```

这个文件中的 mTitle、mDirectors、mActors 为参数，参数使用{{...}}包含，表示在这个位置用参数的值代替{{...}}。

在服务器提交模板文件 movie.html 时可以向该文件传递这些参数。服务器程序 server.py 可以编写为：

```python
import flask
app=flask.Flask(__name__,static_folder="images")
@app.route("/")
def index():
    return flask.render_template("movie.html",mTitle="肖申克的救赎",mDirectors="弗兰克·德拉邦特 Frank Darabont",mActors="蒂姆·罗宾斯 Tim Robbins/...")
app.run()
```

运行该服务器程序，使用浏览器访问"http://127.0.0.1:5000"的结果如图 3-6-2 所示。

图 3-6-2 使用字符串参数的结果

其中，服务器程序中的 flask.render_template()函数向模板文件 movie.html 传递了 mTitle、mDirectors、mActors 参数，因此服务器中如下模板文件 movie.html 的各个参数均被做了替换：

```html
<body>
    名称:{{mTitle}}<br>
    导演:{{mDirectors}}<br>
    主演:{{mActors}}
</body>
```

服务器提交到浏览器端的实际上是以下 HTML 文件：

```
<body>
    名称:肖申克的救赎<br>
    导演:弗兰克·德拉邦特 Frank Darabont<br>
    主演:蒂姆·罗宾斯 Tim Robbins/...
</body>
```

其中,{{mTitle}}、{{mDirectors}}、{{mActors}}都被实际值代替了。

### 2. 模板使用字典参数

前面的 mTitle、mDirectors、mActors 这 3 个参数是分开传递的,如果要传递的参数很多,这样分开传递是不现实的,可以将多个参数打包成一个字典参数传递。服务器程序 server.py 修改为:

```python
import flask
app=flask.Flask(__name__,static_folder="images")
@app.route("/")
def index():
    m={}
    m["mTitle"]="肖申克的救赎"
    m["mDirectors"]="弗兰克·德拉邦特 Frank Darabont"
    m["mActors"]="蒂姆·罗宾斯 Tim Robbins/..."
    return flask.render_template("movie.html",movie=m)
app.run()
```

这个服务器程序向模板文件 movie.html 传递一个 movie 参数,这个参数是一个字典。这时,模板文件 movie.html 相应改成如下形式:

```
<body>
    名称:{{movie["mTitle"]}}<br>
    导演:{{movie["mDirectors"]}}<br>
    主演:{{movie["mActors"]}}
</body>
```

其中,{{movie["mTitle"]}}、{{movie["mDirectors"]}}、{{movie["mActors"]}}就包含电影名称、导演、主演等字典值,效果与前面完全一样。

### 3. 模板使用循环语句

如果要传递很多部电影,可以把这些电影组成一个列表。服务器程序 server.py 如下:

```python
import flask
app=flask.Flask(__name__,static_folder="images")
@app.route("/")
def index():
    movies=[]
    movies.append({"mTitle":"肖申克的救赎","mDirectors":"弗兰克·德拉邦特 Frank Darabont","mActors":"蒂姆·罗宾斯 Tim Robbins/..."})
    movies.append({"mTitle":"这个杀手不太冷","mDirectors":"吕克·贝松 Luc Besson","mActors":"让·雷诺 Jean Reno / 娜塔莉·波特曼 ..."})
    movies.append({"mTitle":"辛德勒的名单","mDirectors":"史蒂文·斯皮尔伯格 Steven Spielberg","mActors":"连姆·尼森 Liam Neeson..."})
    return flask.render_template("movie.html",movies=movies)
app.run()
```

服务器程序向模板文件 movie.html 传递了一个列表 movies，列表中的每个元素都是一个字典，这时模板文件可以使用循环语句来获取 movies 列表的每个元素。循环语句的格式如下：

```
{% for 变量 in 列表变量 %}
...
{% endfor %}
```

相应的模板文件 movie.html 修改如下：

```
<body>
{%for movie in movies %}
  名称：{{movie["mTitle"]}}<br>
  导演：{{movie["mDirectors"]}}<br>
  主演 x：{{movie["mActors"]}}<p>
{% endfor %}
</body>
```

其中，{%for movie in movies %}是 Flask 规定的模板使用的循环语句，这个循环必须使用{% endfor %}标注结束位置，被它们包含的部分就是循环体。movie 表示 movies 列表的每个元素，因此是一个字典，在循环体中通过{{movie["mTitle"]}}、{{movie["mDirectors"]}}、{{ movie["mActors"]}}得到这个字典的值。

运行服务器程序，使用浏览器访问"http://127.0.0.1:5000"的结果如图 3-6-3 所示。

生成的 HTML 代码如下：

图 3-6-3　使用循环语句的结果

```
<body>
  名称：肖申克的救赎<br>
  导演：弗兰克·德拉邦特 Frank Darabont<br>
  主演：蒂姆·罗宾斯 Tim Robbins/...<p>
  名称：这个杀手不太冷<br>
  导演：吕克·贝松 Luc Besson<br>
  主演：让·雷诺 Jean Reno / 娜塔莉·波特曼 ...<p>
  名称：辛德勒的名单<br>
  导演：史蒂文·斯皮尔伯格 Steven Spielberg<br>
  主演：连姆·尼森 Liam Neeson...<p>
</body>
```

循环语句{%for movie in movies %}…{% endfor %}很好地在模板中实现了循环构造 HTML 代码的功能。

### 4. 模板使用判断语句

模板常常要根据不同的情况生成不同的 HTML 代码，这就要用到判断语句。判断语句的格式如下：

```
{% if 条件 %}
...
```

```
{% elif 条件 %}
...
{% else %}
...
{% endif %}
```

例如,服务器程序修改为:

```
import flask
app=flask.Flask(__name__,static_folder="images")
@app.route("/")
def index():
    movies=[]
    movies.append({"mTitle":"肖申克的救赎","mCountry":"美国"})
    movies.append({"mTitle":"这个杀手不太冷","mCountry":"法国"})
    movies.append({"mTitle":"霸王别姬","mCountry":"中国"})
    return flask.render_template("movie.html",movies=movies)
app.run()
```

而模板文件修改为:

```
<body>
{%for movie in movies %}
{% if movie["mCountry"] == "美国" %}
<div style="color:blue">名称:{{movie["mTitle"]}}</div>
{% elif movie["mCountry"] == "中国" %}
<div style="color:red">名称:{{movie["mTitle"]}}</div>
{% else %}
<div style="color:black">名称:{{movie["mTitle"]}}</div>
{% endif %}
{% endfor %}
</body>
```

运行服务器程序后用浏览器访问"http://127.0.0.1:5000"的结果如图 3-6-4 所示。其中,"肖申克的救赎"因为是美国电影名称而显示为蓝色,"这个杀手不太冷"因为是法国电影名称而显示为黑色,而"霸王别姬"因为是中国电影名称而显示为红色。

图 3-6-4　使用判断语句的结果

### 3.6.2　创建翻页电影网站

#### 1. 设计电影网站服务器

为了简化网站的内容,每部电影都只显示其中的图像与名称,其他信息暂时不显示,重点考虑翻页的设计。

为了实现翻页,必须定义一个 pageRowCount 变量,代表一页有几行(每行一部电影),pageCount 表示总页数,pageIndex 表示当前页码。其中,pageIndex=1,2,…,pageCount,当 pageIndex=1 时表示当前是第一页,当 pageIndex=2 时表示当前是第二页,当 pageIndex=pageCount 时表示当前是最后一页。

服务器程序 server.py 如下：

```python
import flask
app=flask.Flask(__name__,static_folder="images")
@app.route("/")
def show():
    pageRowCount=5
    if "pageIndex" in flask.request.values:
        pageIndex=int(flask.request.values.get("pageIndex"))
    else:
        pageIndex=1
    startRow=(pageIndex-1)*pageRowCount
    endRow=pageIndex*pageRowCount
    movies=[]
    try:
        fobj=open("movies.csv","r",encoding="utf-8")
        rows=fobj.readlines()
        count=0
        for row in rows:
            if row.strip("\n").strip()!="":
                count+=1
        count=count-1
        pageCount=count//pageRowCount
        if count % pageRowCount!=0:
            pageCount+=1
        rowIndex=0
        for i in range(1,count+1):
            row=rows[i]
            if rowIndex>=startRow and rowIndex<endRow:
                row=row.strip("\n")
                s=row.split(",")
                m={}
                m["ID"] = s[0]
                m["mImage"] = s[0] + s[1]
                m["mTitle"] = s[2]
                movies.append(m)
            rowIndex+=1
        fobj.close()
    except Exception as err:
        print(err)
    return flask.render_template("movie.html",
movies=movies,pageIndex=pageIndex,pageCount=pageCount)

app.run()
```

程序先从 flask.request.values 中获取当前页码 pageIndex 的值，如果该变量没有值就设置 pageIndex=1 来显示第一页，再根据 pageRowCount 的值与 pageIndex 的值计算出这一页应该显示的行的区间[startRow,endRow]：

```
startRow=(pageIndex-1)*pageRowCount
endRow=pageIndex*pageRowCount
```

打开 movies.csv 文件,读取所有行 rows,计算出有用的行数 count 及 pageCount 的值:

```
count=0
for row in rows:
    if row.strip("\n").strip()!="":
        count+=1
count=count-1
pageCount=count//pageRowCount
if count % pageRowCount!=0:
    pageCount+=1
```

使用一个循环从第二行开始获取每一行(第一行是标题),rowIndex 用于记录行号,只有在 rowIndex>=startRow 同时 rowIndex<endRow 范围内的才是要显示的行,然后获取数据并将其填充到列表 movies 中。

服务器向模板传递的参数有 movies 列表、pageIndex 与 pageCount,其中,pageIndex 与 pageCount 是给模板做翻页按钮使用的。

### 2. 设计电影网站模板

电影网站模板 movie.html 设计如下:

```
<style>
.pic {display:inline-block;width:60px; vertical-align:top;margin:5px;}
.info { display:inline-block; }
.liClass { list-style: none; margin:20px;}
.link {border: 1px solid ;}
a:link { color: blue; text-decoration: none; }
a:visited { color: blue; text-decoration: none; }
</style>
<div>
<ul>
{% for m in movies %}
<li class="liClass"  >
    <div class="pic">
      <img width="50" height="50" src='images/{{m["mImage"]}}'>
   </div>
   <div class="info">
       <div class="title"><h3 style="display:inline-block">
{{m["mTitle"]}}
</h3></div>
    </div>
 </li>
{% endfor %}
</ul>
</div>
<div align="center" class="paging">
    <a href="/?pageIndex=1" class="link">第一页</a>
    {% if pageIndex>1 %}
       <a href="/?pageIndex={{pageIndex-1}}" class="link">前一页</a>
    {% else %}
       <a href="#" class="link">前一页</a>
    {% endif %}
    {% if pageIndex<pageCount %}
       <a href="/?pageIndex={{pageIndex+1}}" class="link">下一页</a>
    {% else %}
```

```html
            <a href="#">下一页</a>
        {% endif %}
        <a href="/?pageIndex={{pageCount}}" class="link">末一页</a>
        <span>Page {{pageIndex}}/{{pageCount}}</span>
</div>
</body>
```

这个模板使用循环语句显示每部电影的图像与名称,各部电影放在一组<ul><li>…</li></ul>中。

模板的重点是翻页的设计,服务器程序向模板文件传递了两个控制翻页的参数,一个是当前页码 pageIndex,另一个是总页数 pageCount,每显示一页都会向服务器程序提交一个参数 pageIndex,服务器程序根据这个参数来组织要显示的数据 movies。

"第一页"的<a>元素使用链接:

```html
<a href="/?pageIndex=1" class="link">第一页</a>
```

向服务器程序传递 pageIndex 参数,于是在服务器程序中使用:

```python
pageIndex=int(flask.request.values.get("pageIndex"))
```

接收到 pageIndex=1,显示第一页的数据。

"末一页"的<a>元素使用链接:

```html
<a href="/?pageIndex={{pageCount}}" class="link">末一页</a>
```

向服务器程序传递 pageIndex 参数,值是 pageCount,于是在服务器程序中接收到 pageIndex=pageCount,显示最后一页的数据。

"下一页"的<a>元素使用下列程序:

```html
{% if pageIndex<pageCount %}
        <a href="/?pageIndex={{pageIndex+1}}" class="link">下一页</a>
    {% else %}
        <a href="#">下一页</a>
{% endif %}
```

当 pageIndex<pageCount 时,下一页的页码是 pageIndex+1。如果 pageIndex=pageCount,就表示到了最后一页,设置超链接 href="#"使得链接不再起作用。

"前一页"的<a>元素使用下列程序:

```html
{% if pageIndex>1 %}
 <a href="/?pageIndex={{pageIndex-1}}" class="link">前一页</a>
   {% else %}
     <a href="#" class="link">前一页</a>
   {% endif %}
```

当 pageIndex>1 时,可以转到前一页的页码 pageIndex-1,不然就设置超链接 href="#"使得链接不再起作用。

这样就设计好服务器程序与模板文件了。运行服务器程序后,使用浏览器访问"http://127.0.0.1:5000",就可以看到图 3-6-1 所示的网页,单击翻页的各个按钮就可以在各个网页之间进行切换。

### 3.6.3 编写爬虫程序

为了使这个爬虫程序只爬取每部电影的名称,根据已经建立的网站编写的爬虫程序

spider.py 如下：

```
import urllib.request
from bs4 import BeautifulSoup

def myFilter(tag):
    if tag.name=="a" and tag.text=="下一页":
        return True
    return False

def spider(url):
    global count
    try:
        print(url)
        resp=urllib.request.urlopen(url)
        html=resp.read().decode()
        soup=BeautifulSoup(html,"lxml")
        lis=soup.find("ul").find_all("li")
        for li in lis:
            div=li.find("div",attrs={"class":"info"})
            mTitle=li.find("div",attrs={"class":"title"}).find("h3").text
            count+=1
            print(count,mTitle)
        #查找翻页按钮
        link=soup.find("div",attrs={"class":"paging"}).find(myFilter)
        href=link["href"].strip()
        if href!="#":
            url=urllib.request.urljoin(url,href)
            spider(url)
    except Exception as err:
        print(err)
count=0
spider("http://127.0.0.1:5000")
```

这个爬虫程序的主要部分是翻页的设计，程序设计 myFilter() 函数查找 "下一页" 的链接 link，如果 link["href"] 的值是 "#"，则说明已经是最后一页，否则 link["href"] 就是下一页的网址，将其转换为绝对地址后再次递归调用 spider() 函数，就可以爬取下一个页面的数据了。

### 3.6.4 执行爬虫程序

执行爬虫程序，结果表明爬虫程序能爬取到所有页面的数据，部分数据如下：

```
http://127.0.0.1:5000
1 肖申克的救赎
2 霸王别姬
3 这个杀手不太冷
4 阿甘正传
5 美丽人生
http://127.0.0.1:5000/?pageIndex=2
6 泰坦尼克号
7 千与千寻
```

```
    8 辛德勒的名单
    9 盗梦空间
    10 机器人总动员
    ……
    http://127.0.0.1:5000/?pageIndex=50
    246 枪火
    247 海蒂和爷爷
    248 叫我第一名
    249 燕尾蝶
    250 穆赫兰道
```

实际上,这个网站的网页虽然多,但是网页的链接是很简单的,各个网页之间通过超链接组成一条直线,是网站有向图中最简单的一种结构。因此,无论是使用递归方法,还是使用深度优先法、广度优先法,程序的执行效率基本上是一样的。

## 3.7 爬取网站全部图像

3-7-A  3-7-B
知识讲解  操作演练

**任务目标**

爬取网站的图像是经常遇到的问题,爬取图像与爬取一般的文字数据有些区别,因为图像数据量一般比较大。爬取小的图像会比较快,爬取大的图像会比较慢,如何高效地爬取这些图像是本节要讨论的问题。在设计的电影网站中,每部电影都有一幅图像,总共有 250 幅图像分布在不同的网页中。在本节中,我们将编写一个高效的爬虫程序来爬取网站的所有图像。

### 3.7.1 创建模拟电影网站

首先设计一个网站,使得网站的图像在传递给客户端的过程中出现不同程度的延迟,以此来模拟实际网络的情况。

#### 1. 设计网站模板

这个网站模板与 3.6 节中的模板相比,最大的区别是把语句:

```
<img width="50" height="50" src='/images/{{m["mImage"]}}'>
```

修改为:

```
<img width="50" height="50" src='/getImage/{{m["mImage"]}}'>
```

图像不再是从网站的静态文件夹 images 中获取,而是从 "/getImage" 路径获取。于是设计模板文件 movie.html 如下:

```
<style>
.pic {display:inline-block;width:60px; vertical-align:top;margin:5px;}
.info { display:inline-block; }
.liClass { list-style: none; margin:20px;}
.link {border: 1px solid ;}
a:link { color: blue; text-decoration: none; }
a:visited { color: blue; text-decoration: none; }
</style>
```

```html
<div>
<ul>
{% for m in movies %}
<li class="liClass"  >
    <div class="pic">
        <img width="50" height="50" src='/getImage/{{m["mImage"]}}'>
   </div>
    <div class="info">
        <div class="title"><h3 style="display:inline-block">
{{m["mTitle"]}}</h3></div>
    </div>
</li>
{% endfor %}
</ul>
</div>
<div align="center" class="paging">
    <a href="/?pageIndex=1" class="link">第一页</a>
    {% if pageIndex>1 %}
       <a href="/?pageIndex={{pageIndex-1}}"  class="link">前一页</a>
    {% else %}
       <a href="#"  class="link">前一页</a>
    {% endif %}
    {% if pageIndex<pageCount %}
       <a href="/?pageIndex={{pageIndex+1}}"  class="link">下一页</a>
    {% else %}
       <a href="#">下一页</a>
    {% endif %}
    <a href="/?pageIndex={{pageCount}}"  class="link">末一页</a>
    <span>Page {{pageIndex}}/{{pageCount}}</span>
</div>
</body>
```

## 2. 设计网站服务器程序

服务器程序 server.py 使用 getImage()函数来向客户端传递图像数据，在获取图像的函数 getImage()时，使用语句 time.sleep(random.random())使传递过程延迟 0～1s，以模拟实际网络传递图像时的网络延迟。服务器程序如下：

```python
import flask
import os
import time
import random
app=flask.Flask(__name__,static_folder="images")
@app.route("/")
def show():
    pageRowCount=5
    if "pageIndex" in flask.request.values:
        pageIndex=int(flask.request.values.get("pageIndex"))
    else:
        pageIndex=1
    startRow=(pageIndex-1)*pageRowCount
    endRow=pageIndex*pageRowCount
```

```
        movies=[]
        try:
            fobj=open("movies.csv","r",encoding="utf-8")
            rows=fobj.readlines()
            count=0
            for row in rows:
                if row.strip("\n").strip()!="":
                    count+=1
            count=count-1
            pageCount=count//pageRowCount
            if count % pageRowCount!=0:
                pageCount+=1
            rowIndex=0
            for i in range(1,count+1):
                row=rows[i]
                if rowIndex>=startRow and rowIndex<endRow:
                    row=row.strip("\n")
                    s=row.split(",")
                    m={}
                    m["ID"] = s[0]
                    m["mImage"] = s[0] + s[1]
                    m["mTitle"] = s[2]
                    movies.append(m)
                rowIndex+=1
            fobj.close()
        except Exception as err:
            print(err)
        return flask.render_template("movie.html",movies=movies,
pageIndex=pageIndex,pageCount=pageCount)

@app.route("/getImage/<name>")
def getImage(name):
    data=b""
    if os.path.exists("images\\"+name):
        f=open("images\\"+name,"rb")
        data=f.read()
        f.close()
    time.sleep(random.random())
    return data

app.run()
```

运行服务器程序后使用浏览器访问网页，可以看到各幅图像的出现有不同程度的延迟。

### 3.7.2 使用单线程程序爬取图像

在前面的章节中已经介绍过图像的下载方法，根据网页结构设计的爬取图像的程序 spider.py 如下：

```
import urllib.request
from bs4 import BeautifulSoup
import datetime

def myFilter(tag):
```

```
            if tag.name=="a" and tag.text=="下一页":
                return True
        return False

    def download(count,src):
        try:
            p=src.rfind("/")
            name=src[p+1:]
            resp=urllib.request.urlopen(src)
            data=resp.read()
            f=open("download\\"+name,"wb")
            f.write(data)
            f.close()
            print(count," download ",name)
        except Exception as err:
            print(err)

    def spider(url):
        global count
        try:
            print(url)
            resp=urllib.request.urlopen(url)
            html=resp.read().decode()
            soup=BeautifulSoup(html,"lxml")
            lis=soup.find("ul").find_all("li")
            for li in lis:
                src = li.find("div", attrs={"class": "pic"}).find("img")["src"]
                src = urllib.request.urljoin(url, src)
                count+=1
                download(count,src)
            #查找翻页按钮
            link=soup.find("div",attrs={"class":"paging"}).find(myFilter)
            href=link["href"].strip()
            if href!="#":
                url=urllib.request.urljoin(url,href)
                spider(url)
        except Exception as err:
            print(err)
    count=0
    start=datetime.datetime.now()
    spider("http://127.0.0.1:5000")
    end=datetime.datetime.now()
    print("总时间(秒):",(end-start).seconds)
```

执行该程序可以爬取全部图像，整个爬取过程耗时130s，部分结果如下：

```
http://127.0.0.1:5000
1  download  000001.jpg
2  download  000002.jpg
3  download  000003.jpg
4  download  000004.jpg
5  download  000005.jpg
......
```

```
http://127.0.0.1:5000/?pageIndex=50
246  download  000246.jpg
247  download  000247.jpg
248  download  000248.jpg
249  download  000249.jpg
250  download  000250.jpg
总时间(秒)：130
```

一般编写的爬虫程序只有一个主线程，这种程序的特点是程序语句是按顺序一条一条执行的，如果上一条语句没有执行完毕，那么下一条语句就没有办法执行。在爬取图像时，如果一幅图像的爬取过程比较慢，就会阻塞程序的进行，直到该图像爬取完后才爬取下一幅图像。这种程序的优点是简单，但是效率不高，如果遇到一幅特别难下载的图像，那么整个程序就会被严重阻塞。

一个比较好的解决方案是采用多线程程序，即为每个下载过程设计一个子线程，让下载图像与爬取文本数据分开，图像下载单独在子线程中执行，这样爬取过程会加快，用户体验会好很多。

关于线程程序，一个形象的比喻是，假设有一个车队要从同一个地方出发，车全部要开到终点。单线程程序就好比整个车队的车在一条单车道的公路行驶。显然，只要道路上有一辆车很慢，就会严重阻碍后面车的行进，即便后面的车有能力开得很快。而多线程程序好比为车队的每辆车都开辟了一条道路，每辆车都尽量快速地开到终点，即便有部分车开得慢也不影响其他的车，因此所有车全部开到终点的时间会比单车道的情况下短很多。

### 3.7.3 使用 Python 的多线程

#### 1. Python 线程类型

在 Python 中，要启动一个线程，可以使用 threading 包中的 Thread 类建立一个对象。Thread 类的基本原型是：

```
t=Thread(target,args=None)
```

其中，target 是要执行的线程函数，args 是一个元组或者列表，作用是为 target()函数提供参数，然后程序调用 t.start()启动线程。

**例 3-7-1**：在主线程中启动一个子线程来执行 reading()函数。

```
import threading
import time
import random

def reading():
    for i in range(10):
        print("reading",i)
        time.sleep(random.randint(1,2))

r=threading.Thread(target=reading)
r.setDaemon(False)
r.start()
print("The End")
```

结果：
```
reading 0
```

```
The End
reading 1
reading 2
reading 3
reading 4
```

从结果可以看出，主线程启动子线程 r 后就结束了，但是子线程还没有结束，显示完 reading 4 后才结束。其中，r.setDaemon(False)用于设置线程 r 为后台线程，后台线程不因主线程的结束而结束。

例 3-7-2：启动一个前台线程。

如果 r 是一个线程，那么设置 r.setDaemon(True)后，r 就是前台线程。

```python
import threading
import time
import random

def reading():
    for i in range(5):
        print("reading",i)
        time.sleep(random.randint(1,2))

r=threading.Thread(target=reading)
r.setDaemon(True)
r.start()
print("The End")
```

结果：

```
reading 0
The End
```

由此可见，主线程结束后子线程也结束，这就是前台线程的特征。

例 3-7-3：前台与后台线程。

```python
import threading
import time
import random

def reading():
    for i in range(5):
        print("reading",i)
        time.sleep(random.randint(1,2))

def test():
    r=threading.Thread(target=reading)
    r.setDaemon(True)
    r.start()
    print("test end")

t=threading.Thread(target=test)
t.setDaemon(False)
t.start()
print("The End")
```

结果：

```
The End
```

```
reading 0
test end
```

由此可见，主线程启动后台子线程 t 后，主线程就结束了，但是 t 还在执行，在 t 中启动前台子线程 r，之后 t 结束，相应的 r 也结束。

#### 2. 线程的等待

在多线程的程序中，往往一个线程（如主线程）要等待其他线程执行完毕后才继续执行，这可以使用 join()函数，使用的方法是：

线程对象.join()

在一个线程代码中执行这条语句，当前的线程就会停止执行，一直等到指定线程对象的线程执行完毕后才继续执行，即这条语句起到阻塞的作用。

**例 3-7-4**：主线程启动一个子线程并等待其结束后才继续执行。

```
import threading
import time
import random

def reading():
    for i in range(5):
        print("reading",i)
        time.sleep(random.randint(1,2))

t=threading.Thread(target=reading)
t.setDaemon(False)
t.start()
t.join()
print("The End")
```

结果：

```
reading 0
reading 1
reading 2
reading 3
reading 4
The End
```

由此可见，主线程一旦启动子线程 t 执行 reading()函数，t.join()函数就阻塞主线程，一直等到子线程 t 执行完毕后才结束 t.join()函数，程序继续执行后显示 The End。

**例 3-7-5**：一个子线程启动另外一个子线程并等待其结束后才继续执行。

```
import threading
import time
import random

def reading():
    for i in range(5):
        print("reading",i)
        time.sleep(random.randint(1,2))

def test():
    r=threading.Thread(target=reading)
    r.setDaemon(True)
```

```
        r.start()
        r.join()
        print("test end")

t=threading.Thread(target=test)
t.setDaemon(False)
t.start()
t.join()
print("The End")
```

结果：

```
reading 0
reading 1
reading 2
reading 3
reading 4
test end
The End
```

由此可见，主线程启动子线程 t 后，t.join()函数会等待线程 t 结束，在 test()函数中再次启动子线程 r，且 r.join()函数会阻塞线程 t，线程 r 执行完毕后结束 r.join()函数，然后显示 test end，之后线程 t 结束，结束 t.join()函数后回到主线程，主线程显示 The End 后结束。

### 3.7.4　使用多线程程序爬取图像

如果把每次调用 download()函数的过程设置成一个子线程，那么下载图像的过程就变成多线程了。程序修改为：

```
count+=1
T=threading.Thread(target=download,args=[count,title,src])
T.setDaemon(False)
T.start()
TS.append(T)
```

其中，为了在程序结束前等待所有的图像下载完毕，这里设计了一个 TS 的全局线程列表变量，它记录每次启动的线程，在程序结束前执行循环：

```
for T in TS:
    T.join()
```

这样就可以保证程序结束前每个子线程都执行完毕，即每幅图像都下载完毕。

根据前面的分析，采用多线程的爬取程序 spider.py 如下：

```
import urllib.request
from bs4 import BeautifulSoup
import datetime
import threading

def myFilter(tag):
    if tag.name=="a" and tag.text=="下一页":
        return True
    return False

def download(count,src):
    try:
        p=src.rfind("/")
```

```python
                name=src[p+1:]
                resp=urllib.request.urlopen(src)
                data=resp.read()
                f=open("download\\"+name,"wb")
                f.write(data)
                f.close()
                print(count," download ",name)
        except Exception as err:
            print(err)

    def spider(url):
        global count,TS
        try:
            print(url)
            resp=urllib.request.urlopen(url)
            html=resp.read().decode()
            soup=BeautifulSoup(html,"lxml")
            lis=soup.find("ul").find_all("li")
            for li in lis:
                src = li.find("div", attrs={"class": "pic"}).find("img")["src"]
                src = urllib.request.urljoin(url, src)
                count+=1
                T=threading.Thread(target=download,args=[count,src])
                T.start()
                TS.append(T)
            #查找翻页按钮
            link=soup.find("div",attrs={"class":"paging"}).find(myFilter)
            href=link["href"].strip()
            if href!="#":
                url=urllib.request.urljoin(url,href)
                spider(url)
        except Exception as err:
            print(err)

count=0
TS=[]
start=datetime.datetime.now()
spider("http://127.0.0.1:5000")
for T in TS:
    T.join()
end=datetime.datetime.now()
print("总时间(秒):",(end-start).seconds)
```

执行该程序,结果如下:

```
http://127.0.0.1:5000
http://127.0.0.1:5000/?pageIndex=2
http://127.0.0.1:5000/?pageIndex=3
http://127.0.0.1:5000/?pageIndex=4
http://127.0.0.1:5000/?pageIndex=5
http://127.0.0.1:5000/?pageIndex=6
19 download 000019.jpg
...
```

```
221 download 000221.jpg
211 download 000211.jpg
223 download 000223.jpg
212 download 000212.jpg
244 download 000244.jpg
总时间(秒): 1
```

由此可见，各幅图像爬取的速度是不同的，下载快的图像先被爬取，下载慢的图像后被爬取，但是总体耗时要比单线程的程序短很多。

## 3.8 综合项目 爬取模拟电影网站数据

3-8-A  3-8-B
知识讲解  操作演练

### 任务目标

首先，根据 movies.csv 文件中的数据，按照真实电影网站建立一个模拟电影网站，如图 3-8-1 所示；然后，编写爬虫程序爬取所有数据与图像，数据被存储到数据库中。本项目将综合使用前面学习到的各种知识与技能，为后面爬取真实的电影网站数据做准备。

图 3-8-1 模拟电影网站

### 3.8.1 创建模拟电影网站

#### 1. 创建网站服务器程序

建立一个电影网站服务器，该服务器读取 movies.csv 文件中的每行数据，并向客户端提交电影列表 movies、当前页码 pageIndex 以及总页数 pageCount 等参数。服务器程序使用 getImage() 函数传递图像，传递过程中使用时间延迟语句来模拟实际的网络传输过程。服务器程序 server.py 设计如下：

```python
import flask
import os
import time
import random

app=flask.Flask(__name__)

@app.route("/")
def show():
    pageRowCount=4
```

```python
            if "pageIndex" in flask.request.values:
                pageIndex=int(flask.request.values.get("pageIndex"))
            else:
                pageIndex=1
        startRow=(pageIndex-1)*pageRowCount
        endRow=pageIndex*pageRowCount

        movies=[]
        try:
            fobj=open("movies.csv","r",encoding="utf-8")
            rows=fobj.readlines()
            count=0
            for row in rows:
                if row.strip("\n").strip()!="":
                    count+=1
            count=count-1
            pageCount=count//pageRowCount
            if count % pageRowCount!=0:
                pageCount+=1
            rowIndex=0

            for i in range(1,count+1):
                row=rows[i]
                if rowIndex>=startRow and rowIndex<endRow:
                    row=row.strip("\n")
                    s=row.split(",")
                    m={}
                    m["ID"] = s[0]
                    m["mImage"] = s[0] + s[1]
                    m["mTitle"] = s[2]
                    m["mNative"] = s[3]
                    m["mNickname"] = s[4]
                    m["mDirectors"] = s[5]
                    m["mActors"] = s[6]
                    m["mTime"] = s[7]
                    m["mCountry"] = s[8]
                    m["mType"] = s[9]
                    m["mPoint"] = s[10]
                    m["mComments"] = s[11]
                    movies.append(m)
                rowIndex+=1
            fobj.close()
        except Exception as err:
            print(err)
        return
    flask.render_template("movie.html",movies=movies,pageIndex=pageIndex,
pageCount=pageCount)

    @app.route("/getImage/<name>")
    def getImage(name):
        data=b""
        if os.path.exists("images\\"+name):
            f=open("images\\"+name,"rb")
```

```
            data=f.read()
            f.close()
        time.sleep(random.random())
        return data

app.run()
```

## 2. 创建网站模板

电影网站模板接收服务器传递的电影列表 movies，使用循环语句显示每部电影的各种信息，同时创建翻页的控制按钮。

网站模板 movie.html 设计如下：

```
<style>
.pic {display:inline-block;width:130px; vertical-align:top;margin:10px;}
.info { display:inline-block; }
.liClass { list-style: none; margin:20px;}
.title { }
.point { margin: 10px;color:red; }
.attrs { margin: 10px;color: #666; }
h3 { display:inline-block;}}
.pl {color:#888;}
.native {}
.nickname {}
.directors {}
.actors {}
.others { fons-size:10px; }
.link {border: 1px solid ;}
a:link { color: blue; text-decoration: none; }
a:visited { color: blue; text-decoration: none; }
</style>
<div>
<ul>
{% for m in movies %}
<li class="liClass" >
    <div class="pic">
      <img width="100"  src='/getImage/{{m["mImage"]}}'>
  </div>
    <div class="info">
        <div class="title"><h3 style="display:inline-block">
{{m["mTitle"]}}</h3>  <span class="point">{{m["mPoint"]}}分</span>
</div>
        <div class="native">
            <span class="pl">原名</span>:<span class="attrs">
{{m["mNative"]}}</span>
        </div>
        <div class="nickname">
            <span class="pl">别名</span>:<span  class="attrs">
{{m["mNickname"]}}</span>
        </div>
        <div class="directors">
            <span class="pl">导演</span>:<span class="attrs">
{{m["mDirectors"]}}</span>
        </div>
```

```
            <div class="actors">
                <span class="pl">主演</span>:<span class="attrs"><span>
{{m["mActors"]}}</span>
            </div>
             <div class="others">
                <span class="pl">时间</span>:<span class="attrs">
{{m["mTime"]}}</span><br>
                <span class="pl">国家</span>:<span class="attrs">
{{m["mCountry"]}}</span><br>
                <span class="pl">类型</span>:<span class="attrs">
{{m["mType"]}}</span><br>
                <span class="pl">评价</span>:<span class="attrs">
{{m["mComments"]}}</span>
            </div>
        </div>
    </li>
      <div></div>
    {% endfor %}
    </ul>
    </div>
    <div align="center" class="paging">
        <a href="/?pageIndex=1" class="link">第一页</a>
        {% if pageIndex>1 %}
          <a href="/?pageIndex={{pageIndex-1}}" class="link">前一页</a>
        {% else %}
          <a href="#" class="link">前一页</a>
        {% endif %}
        {% if pageIndex<pageCount %}
          <a href="/?pageIndex={{pageIndex+1}}" class="link">下一页</a>
        {% else %}
          <a href="#">下一页</a>
        {% endif %}
        <a href="/?pageIndex={{pageCount}}" class="link">末一页</a>
        <span>Page {{pageIndex}}/{{pageCount}}</span>
    </div>
    </body>
```

运行服务器程序后使用浏览器访问"http://127.0.0.1:5000",结果如图 3-8-1 所示。

### 3.8.2 设计存储数据库

设计一个数据库 movies.db 来存储数据,这个数据库中有一张 movies 表,movies 表的各个字段如表 3-8-1 所示。

表 3-8-1 movies 表的各个字段

| 字段名称 | 类型 | 含义 |
| --- | --- | --- |
| ID | varchar(8) | (编号关键字) |
| mTitle | varchar(256) | 名称 |
| mNative | varchar(256) | 原名 |

续表

| 字段名称 | 类型 | 含义 |
|---|---|---|
| mNickname | varchar(256) | 别名 |
| mDirectors | varchar(256) | 导演 |
| mActors | varchar(256) | 主演 |
| mType | varchar(256) | 类型 |
| mTime | varchar(256) | 时间 |
| mCountry | varchar(256) | 国家 |
| mPoint | varchar(8) | 评分 |
| mComments | varchar(256) | 评价 |
| mImage | varchar(8) | 图像扩展名 |

### 3.8.3 编写爬虫程序

根据前面的分析，编写爬虫程序 spider.py，如下：

```python
import urllib.request
from bs4 import BeautifulSoup
import sqlite3
import threading
import os

class MySpider:

    def openDB(self):
        #打开数据库
        self.con = sqlite3.connect("movies.db")
        self.cursor = self.con.cursor()

    def initDB(self):
        #初始化数据库，创建movies表
        try:
            self.cursor.execute("drop table movies")
        except:
            pass
        sql="create table movies (ID varchar(8) primary key,mTitle varchar(256),mNative varchar(256),"
        sql+="mNickname varchar(256),mDirectors varchar(256),mActors varchar(256),mType varchar(256),"
        sql+="mCountry varchar(256),mTime varchar(256),mPoint varchar(8),mComments varchar(256),mImage varchar(256))"
        self.cursor.execute(sql)
        self.count=0
        self.TS=[]

    def closeDB(self):
        #关闭数据库
```

```python
            self.con.commit()
            self.con.close()
    def insertDB(self,ID,mTitle,mNative,mNickname ,mDirectors ,
mActors ,mType ,mCountry ,mTime ,mPoint,mComments,mImage):
        #插入一条记录
        try:
            sql="insert into movies (ID,mTitle,mNative,mNickname ,
mDirectors ,mActors ,mType ,mCountry ,mTime ,mPoint,"
            sql+="mComments,mImage) values (?,?,?,?,?,?,?,?,?,?,?,?)"
            self.cursor.execute(sql,[ID,mTitle,mNative,mNickname ,
mDirectors ,mActors ,mType ,mCountry ,mTime ,mPoint,mComments,mImage])
        except Exception as err:
            print(err)

    def download(self, imgName,src):
        #下载图像并将其存储到download文件夹
        try:
            resp = urllib.request.urlopen(src)
            data = resp.read()
            f = open("download\\" + imgName, "wb")
            f.write(data)
            f.close()
            print("download ", imgName)
        except Exception as err:
            print(err)

    def myFilter(self,tag):
        #过滤函数
        if tag.name == "a" and tag.text == "下一页":
            return True
        return False

    def spider(self,url):
        #爬虫函数
        try:
            resp = urllib.request.urlopen(url)
            html = resp.read().decode()
            soup = BeautifulSoup(html, "lxml")
            #获取所用的<li>
            lis = soup.find("ul").find_all("li")
            for li in lis:
                #爬取电影名称等数据
                div = li.find("div", attrs={"class": "info"})
                mTitle = div.find("div", attrs={"class": "title"}).find("h3").text
                mPoint = div.find("div", attrs={"class": "title"}).find("span",attrs={"class":"point"}).text
                mNative = div.find("div", attrs={"class": "native"}).find("span", attrs={"class": "attrs"}).text
                mNickname = div.find("div", attrs={"class": "nickname"}).find("span",attrs={"class":"attrs"}).text
```

```python
                        mDirectors = div.find("div", attrs={"class": "directors"}).find("span",attrs={"class":"attrs"}).text
                        mActors = div.find("div", attrs={"class": "actors"}).find("span", attrs={"class": "attrs"}).text
                        spans = div.find("div", attrs={"class": "others"}).find_all("span", attrs={"class": "attrs"})
                        mTime = spans[0].text
                        mCountry = spans[1].text
                        mType = spans[2].text
                        mComments = spans[3].text
                        src = li.find("div", attrs={"class": "pic"}).find("img")["src"]
                        src = urllib.request.urljoin(url, src)
                        #构造ID
                        self.count += 1
                        ID=str(self.count)
                        while len(ID)<6:
                            ID="0"+ID
                        p=src.rfind(".")
                        mImage=src[p:]
                        #启动子线程来下载图像
                        T = threading.Thread(target=self.download, args=[ID+mImage, src])
                        T.setDaemon(False)
                        T.start()
                        self.TS.append(T)
                        print(ID,mTitle)
                        self.insertDB(ID,mTitle,mNative,mNickname ,mDirectors ,mActors ,mType ,mCountry ,mTime ,mPoint,mComments,mImage)
                #查找翻页按钮
                link = soup.find("div", attrs={"class": "paging"}).find(self.myFilter)
                href = link["href"].strip()
                if href != "#":
                    url = urllib.request.urljoin(url, href)
                    self.spider(url)
        except Exception as err:
            print("spider: " + str(err))

    def process(self):
        #爬取过程函数
        if not os.path.exists("download"):
            os.mkdir("download")
        self.openDB()
        self.initDB()
        self.spider("http://127.0.0.1:5000")
        self.closeDB()
        #在爬取结束前等待各个子线程结束，保证各幅图像完整下载
        for T in self.TS:
            T.join()

    def show(self):
```

```
            #显示数据库数据
            self.openDB()
            self.cursor.execute("select ID,mImage,mTitle,mNative,
mNickname ,mDirectors ,mActors ,mType ,mCountry ,mTime ,mPoint,mComments from
movies")
            rows=self.cursor.fetchall()
            for row in rows:
                for r in row:
                    print(r)
                print()
            self.closeDB()
            print("Total ",len(rows))

#主程序
spider=MySpider()
while True:
    print("1.爬取")
    print("2.显示")
    print("3.退出")
    s=input("选择(1,2,3):")
    if s=="1":
        spider.process()
    elif s=="2":
        spider.show()
    elif s=="3":
        break
```

### 3.8.4 执行爬虫程序

先执行服务器程序,再执行爬虫程序,可以爬取所有的电影数据并将其存储到数据库,同时爬取所有的电影图像并将其存储到 download 文件夹,每部电影的图像用它的 ID 识别。例如,ID="000001"的电影图像是"000001.jpg"。下面是爬取到的部分电影数据:

```
000001
.jpg
肖申克的救赎
The Shawshank Redemption
月黑高飞  /  刺激1995
弗兰克·德拉邦特 Frank Darabont
蒂姆·罗宾斯 Tim Robbins /...
犯罪 剧情
美国
1994
9.6分
1241123人


000002
.jpg
霸王别姬
Farewell My Concubine
```

陈凯歌 Kaige Chen
张国荣 Leslie Cheung / 张丰毅 Fengyi Zha...
剧情 爱情
中国
1993
9.6 分
914326 人

...
000250
.jpg
穆赫兰道
Mulholland Dr.
穆荷兰大道　/　失忆大道
大卫·林奇 David Lynch
娜奥米·沃茨 Naomi Watts / 劳拉·哈...
剧情 悬疑 惊悚
法国 美国
2001
8.3 分
266223 人

## 3.9　实战项目　爬取实际电影网站数据

3-9-A　3-9-B

知识讲解　操作演练

**任务目标**

在本实战项目中，要求设计一个爬虫程序，爬取"豆瓣电影 Top 250"中总共 250 部电影的信息。

### 3.9.1　解析电影网站的 HTML 代码

使用 Chrome 浏览器访问豆瓣电影网站，查看电影列表，单击鼠标右键，在弹出的快捷菜单中选择"检查"命令就可以看到网站的 HTML 代码，如图 3-9-1 所示。

从 HTML 代码中可以看到电影信息在<ol class="grid_view">的各个<li>中，每个<li>中都包含一部电影的信息。复制第一个<li>的 HTML 代码并使用 BeautifulSoup 的 prettify()函数整理后得到：

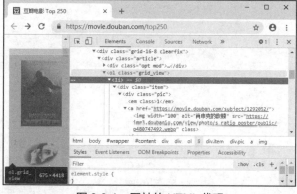

图 3-9-1　网站的 HTML 代码

```
<html>
 <body>
```

```html
<li>
  <div class="item">
    <div class="pic">
     <em class="">
      1
     </em>
      <a href="https://movie.douban.com/subject/1292052/">
        <img alt="肖申克的救赎" class="" src="https://img3.doubanio.com/view/photo/s_ratio_poster/public/p480747492.webp" width="100"/>
      </a>
    </div>
    <div class="info">
     <div class="hd">
      <a class="" href="https://movie.douban.com/subject/1292052/">
       <span class="title">
         肖申克的救赎
       </span>
       <span class="title">
        / The Shawshank Redemption
       </span>
       <span class="other">
        / 月黑高飞  /  刺激1995
       </span>
      </a>
      <span class="playable">
       [可播放]
      </span>
     </div>
     <div class="bd">
      <p class="">
       导演: 弗兰克·德拉邦特 Frank Darabont    主演: 蒂姆·罗宾斯 Tim Robbins /...
       <br/>
       1994 / 美国 / 犯罪 剧情
      </p>
      <div class="star">
       <span class="rating5-t">
       </span>
       <span class="rating_num" property="v:average">
        9.6
       </span>
       <span content="10.0" property="v:best">
       </span>
       <span>
        1260117人评价
       </span>
      </div>
      <p class="quote">
       <span class="inq">
        希望让人自由。
       </span>
      </p>
```

```
            </div>
        </div>
    </div>
  </li>
 </body>
</html>
```

### 3.9.2 爬取电影网站数据

现在有两个页面展示了电影信息,接下来爬取这些信息数据。

**1. 爬取电影名称**

从 HTML 代码中很容易找到<div class="info">下面的两个<span class="title">,从而得到电影名称 mTitle 与原名 mNative,在<span class="other">中得到别名 mNickname。爬取程序如下:

```
div = li.find("div", attrs={"class": "info"})
hd = div.find("div", attrs={"class": "hd"})
spans = hd.find_all("span", attrs={"class": "title"})
mTitle = spans[0].text.replace("\n","").strip()
mNative = spans[1].text.replace("\n","").strip() if len(spans)>1 else ""
mNickname=hd.find("span", attrs={"class": "other"}).text.replace("\n","").strip()
```

**2. 爬取评分与评价人数**

从 HTML 代码中可以看到<div class="star">中的<span class="rating_num">,它下面就是评分。另外,最后一个<li>中的内容就是评价人数。因此评分 mPoint 与评价人数 mComment 可以通过下列程序爬取:

```
sdiv = li.find("div", attrs={"class": "star"})
mPoint = sdiv.find("span", attrs={"class": "rating_num"}).text.replace("\n","").strip()
mComment = sdiv.find_all("span")[-1].text.replace("\n","").strip()
```

**3. 爬取其他信息**

在 HTML 代码的<div class="bd">下面的第一个<p>中有导演 mDirectors、主演 mActors、时间 mTime、国家 mCountry、类型 mType 等数据,例如:

```
<p class="">
    导演: 弗兰克·德拉邦特 Frank Darabont    主演: 蒂姆·罗宾斯 Tim Robbins /...
    <br/>
    1994 / 美国 / 犯罪 剧情
</p>
```

这些数据可以通过 splitItems()函数得到:

```
def splitItems(p):
    res = []
    flag=True
    for c in p.children:
        if isinstance(c, bs4.element.NavigableString):
            t = c.string.replace(" ", "").replace("\n", "")
            if t!="":
                if flag:
```

```
                        pos = t.find("主演")
                        director = t[:pos].replace("导演:","")
                        actor = t[pos + 3:]
                        res.append(director)
                        res.append(actor)
                    else:
                        st=t.split("/")
                        for e in st:
                            res.append(e)
                        break
            elif isinstance(c, bs4.element.Tag) and c.name=="br":
                flag=False
    return res
```

splitItems()函数用于查找<p>的所有孩子节点，在找到<br>之前如果找到一个非空的bs4.element.NavigableString，就找到了导演与主演部分。例如，"导演: 弗兰克·德拉邦特 Frank Darabont  主演: 蒂姆·罗宾斯 Tim Robbins /..."，从中分离出导演与主演。类似的，可从"1994 / 美国 / 犯罪 剧情"中分离出时间、国家、类型。最后函数返回一个列表，依次是导演、主演、时间、国家、类型。

```
res=self.splitItems(p)
mDirectors=res[0] if len(res)>0 else ""
mActors=res[1] if len(res)>1 else ""
mTime=res[2] if len(res)>2 else ""
mCountry=res[3] if len(res)>3 else ""
mType=res[4] if len(res)>4 else ""
```

### 4．爬取电影图像

电影图像存储在<img src=...>指定的 src 地址中，因此可以使用下列语句获取电影图像的 src 地址：

```
img = li.find("div", attrs={"class": "pic"}).find("img")
src=urllib.request.urljoin(url,img["src"])
```

获取 src 地址后，使用子线程启动下载函数 download()来爬取电影图像。

### 5．实现网页翻页

在网页的底部找到翻页按钮，查看它们的 HTML 代码，如图 3-9-2 所示。

复制翻页的 HTML 代码，整理后得到：

图 3-9-2　网页翻页的 HTML 代码

```
<div class="paginator">
        <span class="prev">&lt;前页</span>
        <span class="thispage">1</span>
            <a href="?start=25&amp;filter=">2</a>
            <a href="?start=50&amp;filter=">3</a>
            <a href="?start=75&amp;filter=">4</a>
            <a href="?start=100&amp;filter=">5</a>
            <a href="?start=125&amp;filter=">6</a>
            <a href="?start=150&amp;filter=">7</a>
```

```
            <a href="?start=175&amp;filter=">8</a>
            <a href="?start=200&amp;filter=">9</a>
            <a href="?start=225&amp;filter=">10</a>
        <span class="next">
            <link rel="next" href="?start=25&amp;filter=">
            <a href="?start=25&amp;filter=">后页&gt;</a>
        </span>
        <span class="count">(共250条)</span>
</div>
```

从上述代码可以看到,只要找到<div class="paginator">中<span class="next">中的<a>就能爬取下一页的链接,然后得到下一页的地址,再次递归调用爬虫程序就可以实现翻页爬取,程序如下:

```python
div = soup.find("div", attrs={"class": "paginator"})
link = div.find("span", attrs={"class": "next"}).find("a")
if link:
    href = link["href"]
    url = urllib.request.urljoin(url, href)
    # 递归调用爬虫程序
```

### 3.9.3 编写爬虫程序

设计一个数据库 movies.db 来存储数据,这个数据库中有一张 movies 表,表的结构与表 3-8-1 中设计的完全一样。

根据前面的分析,编写爬虫程序 spider.py 如下:

```python
import urllib.request
from bs4 import BeautifulSoup
import sqlite3
import threading
import bs4
import os
import random

#用于模拟HTTP头的User-Agent
ua_list = [
        "Mozilla/5.0 (Macintosh; Intel Mac OS X 10.6; rv2.0.1) Gecko/20100101 Firefox/4.0.1",
        "Mozilla/5.0 (Windows NT 6.1; rv2.0.1) Gecko/20100101 Firefox/4.0.1",
        "Opera/9.80 (Macintosh; Intel Mac OS X 10.6.8; U; en) Presto/2.8.131 Version/11.11",
        "Opera/9.80 (Windows NT 6.1; U; en) Presto/2.8.131 Version/11.11",
        "Mozilla/5.0 (Macintosh; Intel Mac OS X 10_7_0) AppleWebKit/535.11 (KHTML, like Gecko) Chrome/17.0.963.56 Safari/535.11"
        ]

class MySpider:

    def openDB(self):
        #打开数据库
```

```python
        self.con = sqlite3.connect("movies.db")
        self.cursor = self.con.cursor()

    def initDB(self):
        #初始化数据库,创建movies表
        try:
            self.cursor.execute("drop table movies")
        except:
            pass
        self.cursor.execute("create table movies (ID varchar(8) primary key,mTitle varchar(256),mNative,mNickname varchar(256),mDirectors varchar(256),mActors varchar(256),mTime varchar(256),mCountry varchar(256),mType varchar(256),mPoint varchar(256),mComment varchar(256))")
        self.count=0
        self.TS=[]

    def closeDB(self):
        #关闭数据库
        self.con.commit()
        self.con.close()

    def insertDB(self,ID,mTitle,mNative,mNickname ,mDirectors ,mActors ,mTime,mCountry ,mType,mPoint,mComment):
        #插入一条记录
        try:
            sql="insert into movies (ID,mTitle,mNative,mNickname ,mDirectors ,mActors ,mTime,mCountry ,mType,mPoint,mComment) values (?,?,?,?,?,?,?,?,?,?,?)"
            self.cursor.execute(sql,[ID,mTitle,mNative,mNickname ,mDirectors ,mActors ,mTime,mCountry ,mType,mPoint,mComment])
        except Exception as err:
            print(err)

    def download(self, ID,src):
        #下载图像并将其保存到download文件夹
        try:
            req=urllib.request.Request(url=src,headers={"User-Agent": random.choice(ua_list)})
            resp = urllib.request.urlopen(req)
            data = resp.read()
            f = open("download\\" + ID + ".jpg", "wb")
            f.write(data)
            f.close()
            print("download ", ID+".jpg")
        except Exception as err:
            print(err)

    def splitItems(self,p):
        #分解每个项目字段
        res = []
        flag = True
        for c in p.children:
```

```python
                    if isinstance(c, bs4.element.NavigableString):
                        t = c.string.replace("\n", "").strip()
                        if t != "":
                            if flag:
                                pos = t.find("主演")
                                director = t[:pos].replace("导演:", "")
                                actor = t[pos + 3:]
                                res.append(director.strip())
                                res.append(actor.strip())
                            else:
                                st = t.split("/")
                                for e in st:
                                    res.append(e.strip())
                                break
                    elif isinstance(c, bs4.element.Tag) and c.name == "br":
                        flag = False
            return res

    def spider(self,url):
        #爬虫函数
        try:
            print(url)
            req=urllib.request.Request(url=url,headers={"User-Agent": random.choice(ua_list)})
            resp = urllib.request.urlopen(req)
            html = resp.read().decode()
            soup = BeautifulSoup(html, "lxml")
            #获取所用的<li>
            lis = soup.find("div", attrs={"id": "content"}).find("ol", attrs={"class": "grid_view"}).find_all("li")
            for li in lis:
                #爬取电影名称等数据
                div = li.find("div", attrs={"class": "info"})
                hd=div.find("div", attrs={"class": "hd"})
                spans =hd.find_all("span", attrs={"class": "title"})
                mTitle = spans[0].text.replace("\n","").strip() if len(spans)>0 else ""
                mNative = spans[1].text.replace("\n","").strip() if len(spans)>1 else ""
                mNickname=hd.find("span", attrs={"class": "other"}).text.replace("\n","").strip()
                sdiv = li.find("div", attrs={"class": "star"})
                mPoint = sdiv.find("span", attrs={"class": "rating_num"}).text.replace("\n","").strip()
                mComment = sdiv.find_all("span")[-1].text.replace("\n","").strip()
                bd=div.find("div", attrs={"class": "bd"})
                p=bd.find("p")
                #下面几个数据需要分解
                res=self.splitItems(p)
                mDirectors=res[0] if len(res)>0 else ""
                mActors=res[1] if len(res)>1 else ""
```

```python
                            mTime=res[2] if len(res)>2 else ""
                            mCountry=res[3] if len(res)>3 else ""
                            mType=res[4] if len(res)>4 else ""
                            img = li.find("div", attrs={"class": "pic"}).find("img")
                            src=urllib.request.urljoin(url,img["src"])
                            #创建 ID
                            self.count += 1
                            ID=str(self.count)
                            while len(ID)<6:
                                ID="0"+ID
                            #启动子线程下载图像
                            T = threading.Thread(target=self.download, args=[ID, src])
                            T.setDaemon(False)
                            T.start()
                            self.TS.append(T)
                            self.insertDB(ID,mTitle,mNative,mNickname ,mDirectors ,mActors ,mTime,mCountry ,mType,mPoint,mComment)
                    #网页翻页
                    div = soup.find("div", attrs={"class": "paginator"})
                    link = div.find("span", attrs={"class": "next"}).find("a")
                    if link:
                        href = link["href"]
                        url = urllib.request.urljoin(url, href)
                        #递归调用 spider()
                        self.spider(url)
            except Exception as err:
                print("spider: " + str(err))

    def process(self):
        #爬取过程函数
        if not os.path.exists("download"):
            os.mkdir("download")
        self.openDB()
        self.initDB()
        self.spider("https://movie.douban.com/top250")
        self.show()
        self.closeDB()
        #在程序结束前等待各个子线程,确保完成各幅图像的下载
        for T in self.TS:
            T.join()

    def show(self):
        #显示数据库数据
        self.cursor.execute("select ID,mTitle,mNative,mNickname ,mDirectors ,mActors ,mTime,mCountry ,mType,mPoint,mComment from movies")
        rows=self.cursor.fetchall()
        for row in rows:
            for r in row:
                print(r)
```

```
            print()
            print("Total ",len(rows))

#主程序
spider=MySpider()
while True:
    print("1.爬取")
    print("2.显示")
    print("3.退出")
    s=input("选择(1,2,3):")
    if s=="1":
        spider.process()
    elif s=="2":
        spider.openDB()
        spider.show()
        spider.closeDB()
    elif s=="3":
        break
```

这个程序使用一组随机的 User-Agent 值，每次在访问网站时都会随机取出一个 User-Agent 值并放在 HTTP 的头部，这样做的好处是尽可能地让爬虫程序对 Web 服务器的访问看起来像是对真实浏览器的访问，以此来"骗过"Web 服务器，让爬虫程序能连续地访问 Web 服务器。

### 3.9.4 执行爬虫程序

执行爬虫程序，成功爬取到了 250 部电影的数据，部分数据如下：

```
000001
肖申克的救赎
/ The Shawshank Redemption
/ 月黑高飞 / 刺激 1995
弗兰克·德拉邦特 Frank Darabont
蒂姆·罗宾斯 Tim Robbins /...
1994
美国
犯罪 剧情
9.6
1274771 人评价

000002
霸王别姬
Farewell My Concubine
陈凯歌 Kaige Chen
张国荣 Leslie Cheung / 张丰毅 Fengyi Zha...
1993
中国
剧情 爱情
```

```
9.6
940573 人评价

...
000250
攻壳机动队
/ 攻殼機動隊
/ Ghost in the Shell
押井守 Mamoru Oshii
田中敦子 Atsuko Tanaka / 大冢明夫 Akio...
1995
日本
动作 科幻 动画
8.9
71698 人评价
```

## 项目总结

本项目涉及一个有多个页面的电影网站，我们使用递归、深度优先、广度优先等方法爬取各个网页的数据，实现了爬取电影网站数据的爬虫程序。

递归、深度优先、广度优先等是爬虫程序经常使用的方法。一般的程序都是单线程程序，网页是按一定的顺序一个一个地被爬取的，如果某个网页爬取中出现了延迟，必然会影响后续网页的爬取。一个解决方法是采用多线程，把不同网页的爬取分配到多个不同的线程中，本项目通过爬取网页的图像演示了这种多线程技术。在后面的项目中将进一步讲解分布式爬取技术，scrapy 框架就是一种优秀的分布式爬虫框架。

## 练习 3

1. 什么是深度优先法和广度优先法，它们有什么特点？
2. 如何启动一个 Python 线程？为什么说爬虫程序一般都会使用多线程？
3. 设计两个网页 A.html 与 B.html，它们都包含相同结构的学生信息（姓名、性别、年龄）。A.html 文件与 B.html 文件如下。

（1）A.html：

```
<div>姓名: A1</div><div>性别: 男</div><div>年龄: 20</div>
...
<div>姓名: A10</div><div>性别: 女</div><div>年龄: 18</div>
```

（2）B.html：

```
<div>姓名: B1</div><div>性别: 男</div><div>年龄: 20</div>
...
<div>姓名: B10</div><div>性别: 女</div><div>年龄: 18</div>
```

设计一个包含两个线程的爬虫程序，一个线程爬取 A.html，另一个线程爬取 B.html，爬取的学生信息都保存到列表 students 中，最后显示爬取的结果。

4. 设计一个爬虫程序，爬取一个网站的所有图像。

# 项目 ❹ 爬取图书网站数据

Scrapy 是常用的 Python 爬虫框架之一,它对数据爬取与数据存储进行了分工,实现了数据的分布式爬取与存储,是一个功能强大的专业爬取框架。本项目以爬取当当网的图书数据为例,使用 Scrapy 设计爬虫程序,快速地爬取几千本图书的数据。

使用本项目的程序爬取中国古诗词类书籍,中国是诗的国度,唐诗、宋词、元曲等构成了诗歌国度中无比壮丽的景象,饱含着丰富的文化内涵和审美意蕴,是我们祖先智慧的结晶,是中国文化最灿烂的瑰宝之一。

拓展阅读

经典唐诗 10 首

## 4.1 项目任务

使用浏览器访问当当网,输入"python"关键字进行搜索,可以看到有大量的 Python 类别的图书,如图 4-1-1 所示。由于图书比较多,因此图书信息被分成了几十个页面显示,单击">"按钮可以进入下一页。这个项目的任务就是设计一个 scrapy 爬虫程序,爬取所有页面的图书信息。

4-1-A
知识讲解

4-1-B
操作演练

图 4-1-1 搜索 Python 类别的图书

在爬取实际图书网站的数据之前先爬取一个模拟图书网站的数据。创建项目文件夹 project4,在该文件夹中有一个 books.csv 文件,它包含很多图书的信息,前面几行如下:

ID,bTitle,bAuthor,bDate,bPublisher,bPrice,bDetail,bExt
000001, Python 编程 从入门到实践,埃里克•马瑟斯,人民邮电出版社,2016-07-01,¥62.00,上到有编程基础的程序员,下到 10 岁少年,想入门 Python 并达到可以开发实际项目的水平。本书是一本全面的从入门到实践的 Python 编程教程,可以带领读者快速掌握编程基础知识编写能解决实际问题的代码并开发复杂项目。书中内容分为基础篇和实战篇两部分。基础篇介绍基本的编程概念，如列表、字典、类和循环，并指导读者编写整洁且易于理解的代码。另外还介绍如何让程序能够与用户交互，以及如何在代码运行前进行测试。实战篇介绍如何利用新学到的知识开发功能丰富的项目：2D游戏《外星人入侵》、数据可视化实战、Web应用程序。,,.jpg

000002,Python 编程（第四版）,Mark,中国电力出版社,2014-12-01,¥136.60,重磅推荐：Python 袖珍指南（第五版） 本书是 Python 应用的手册指南，它涵盖了 Python 编程的方方面面，从系统管理到图形界面编程，从文本处理到网络编程，从数据库到语言扩展。在这些主题的探讨中，作者提供了大量的实际代码，通过对这些代码的研读，将提高读者的 Python 编程水平以及实际业务问题的解决能力。,,.jpg

......

文件中的各个数据字段使用","分隔，第一行是各个数据字段的名称，其中 ID 是编号、bTitle 是名称、bAuthor 是作者、bDate 是出版时间、bPublisher 是出版社、bPrice 是价格、bDetail 是图书简介、bExt 是图书的图像扩展名。另外，images 文件夹中有每本图书的封面图像（后文简称图书图像），图像的名称是以图书编号命名的，如图 4-1-2 所示。

图 4-1-2　图书的封面图像

这里使用 books.csv 文件中的数据编写一个模拟图书网站，如图 4-1-3 所示。这个网站显示 480 本图书的信息，信息分为很多个页面显示，单击每个页面的"第一页""前一页""下一页""末一页"等按钮就可以在各个页面之间切换。基于这个网站编写一个 scrapy 爬虫程序，爬取网站的所有图书信息。

图 4-1-3　模拟图书网站

## 4.2 使用 scrapy 创建爬虫程序

4-2-A    4-2-B

知识讲解    操作演练

 任务目标

安装 scrapy 框架，创建一个简单的 Web 服务器，使用 scrapy 编写爬虫程序爬取网站的数据。在本节中，我们主要了解 scrapy 项目的框架结构。

### 4.2.1 创建网站服务器程序

简单起见，使用 Flask 设计服务器程序 server.py：

```
import flask
app=flask.Flask(__name__,static_folder="images")
@app.route("/")
def index():
    return "测试 scrapy"
app.run()
```

运行这个服务器程序，使用浏览器访问"http://127.0.0.1:5000"，就会看到图 4-2-1 所示的 Web 网页。

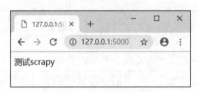

图 4-2-1　Web 网页

### 4.2.2 安装 scrapy 框架

在命令行窗体中执行如下命令完成 scrapy 框架的安装。

```
pip install scrapy
```

### 4.2.3 scrapy 项目的创建

**1. 创建 scrapy 项目**

创建 scrapy 项目的操作步骤如下。

（1）进入命令行窗口，进入 project4 文件夹后执行命令：

```
scrapy startproject demo
```

执行该命令后会创建一个名称为 demo 的 scrapy 项目，如图 4-2-2 所示。

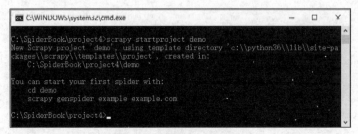

图 4-2-2　创建 scrapy 项目

（2）scrapy 项目创建后会在 project4 中创建 demo 文件夹，demo 文件夹中包含另外一个同名的 demo 子文件夹，demo 子文件夹中包含一个 spiders 子文件夹。scrapy 的项目结构如图 4-2-3 所示。

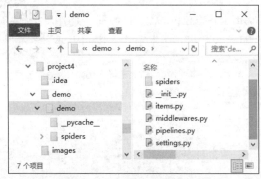

图 4-2-3　scrapy 的项目结构

在 project4\demo\demo 文件夹中有几个重要的文件或文件夹，表 4-2-1 列出了它们的基本作用，之后还会详细介绍它们的功能。

表 4-2-1　scrapy 文件或文件夹的基本作用

| 文件或文件夹名称 | 说明 |
| --- | --- |
| items.py | 设计数据字段的文件，类似定义数据库的字段 |
| middlewares.py | 一个中间件文件，用来设计下载的中间过程 |
| pipelines.py | 一个数据管道文件，爬取的数据被推送到这个管道文件进行存储 |
| settings.py | 一个设置文件，使用它配置爬虫程序的功能 |
| spiders | 一个子文件夹，编写的爬虫程序就保存在这个文件夹中 |

（3）用 PyCharm 打开 project4 文件夹，它包含 demo 文件夹。

（4）在 project4\demo\demo\spiders 文件夹中建立一个自己的 Python 程序，如 MySpider.py，这就是爬虫程序了，程序如下：

```
import scrapy
class MySpider(scrapy.Spider):
    name = "mySpider"
    def start_requests(self):
        url ='http://127.0.0.1:5000'
        yield scrapy.Request(url=url,callback=self.parse)

    def parse(self, response):
        print(response.url)
        data=response.body.decode()
        print(data)
```

（5）在 project4\demo\demo 文件夹中建立一个执行程序，如 run.py，程序如下：

```
from scrapy import cmdline
cmdline.execute("scrapy crawl mySpider -s LOG_ENABLED=False".split())
```

（6）保存这些程序并运行 run.py，可以在 PyCharm 中看到执行的结果：

```
http://127.0.0.1:5000
```

## 项目 ❹ 爬取图书网站数据

测试 scrapy

由此可见,程序 MySpider.py 访问了我们自己的 Web 网站并获取了网站的网页。

### 2. 解析 scrapy 项目

这个项目初步看起来有点复杂,但是仔细分析也不难理解。下面来分析 MySpider.py 程序。

(1)语句:

```
import scrapy
```

引入 scrapy 程序包,这个包中有一个请求对象 Request 类与一个响应对象 Response 类。

(2)语句:

```
class MySpider(scrapy.Spider):
    name = "mySpider"
```

任何一个爬虫程序类都继承 scrapy.Spider 类,任何一个爬虫程序都有一个名字,这个名字在整个爬虫项目中是唯一的,这里把这个爬虫程序的名字定为"mySpider"。

(3)语句:

```
def start_requests(self):
    url ='http://127.0.0.1:5000'
    yield scrapy.Request(url=url,callback=self.parse)
```

url 是爬虫程序的入口地址,start_requests()函数是爬虫程序的入口函数。程序开始时确定要爬取的网站地址,然后建立一个 scrapy.Request 请求类,向这个类提供 url 参数,指明要爬取的网站地址,爬取完成后就执行默认的回调函数 parse()。

值得指出的是,scrapy 的执行过程是异步进行的,即指定一个网址爬取数据时,程序不用一直等待这个网站的响应。显然,如果要等待网站响应,假定网站迟迟不响应,那么整个程序岂不是"卡死了"! scrapy 没有这么设计,它提供一个回调函数机制,爬取网站时同时提供一个回调函数,当网站响应后就触发执行这个回调函数,网站什么时候响应就什么时候调用这个回调函数,这样就可以应对响应时间很长的网站了,这种方法也被称为异步程序方法。

(4)语句:

```
def parse(self, response):
    print(response.url)
    data=response.body.decode()
    print(data)
```

回调函数 parse()包含一个 scrapy.Response 类的对象 response,它包含网站响应的一切信息,其中 response.url 是网站的网址,response.body 是网站相应的二进制数据,即网页的内容。该内容通过 decode()函数解码后变成字符串,然后就可以输出了。

**注意**:MySpider 只是一个类,不能单独执行,要执行这个爬虫程序必须使用 scrapy 中专门的命令 scrapy crawl。

设计一个 Python 程序 run.py,它包含执行命令行的语句:

```
from scrapy import cmdline
cmdline.execute("scrapy crawl mySpider -s LOG_ENABLED=False".split())
```

直接执行 run.py 就可以执行这个爬虫程序了,结果显示在 PyCharm 中。

### 4.2.4 入口函数与入口地址

#### 1. 入口函数

程序中使用了入口函数:

```
def start_requests(self):
    url ='http://127.0.0.1:5000'
    yield scrapy.Request(url=url,callback=self.parse)
```

#### 2. 入口地址

实际上,这个函数也可以用 start_urls 的入口地址来代替:

```
start_urls=['http://127.0.0.1:5000']
```

入口地址可以有多个,因此 start_urls 是一个列表。

入口函数与入口地址的作用是一样的,都是引导函数的开始,实际使用中选择其中一种就可以了。

### 4.2.5 Python 的 yield 语句

在入口函数中有一条 yield 语句,yield 语句是 Python 的一种特殊语句,主要作用是提供一个等待被取走的值。

先看一个实例。

```
def fun():
    s=['a','b','c']
    for x in s:
        yield x
    print("fun End")

f=fun()
print(f)
for e in f:
    print(e)
```

执行该程序,结果如下:

```
<generator object fun at 0x0000003EA99BD728>
a
b
c
fun End
```

由此可见,fun()函数返回一个 generator 对象,这个对象包含一系列的元素,可以使用 for 循环提取,执行循环的过程如下:

```
for e in f:
    print(e)
```

第一次 for 循环,f 执行到 yield 语句,就返回值 a,for 循环从 f 提取的元素是 a,然后 e='a',输出 a。fun()执行到 yield 语句时会等待返回的值被取走,同时 fun()停留在 yield 语句,一旦 yield 语句返回的值被取走,就再次执行循环,yield 语句返回 b。

第二次 for 循环,提取 f 函数的 b 元素,输出 b,然后循环继续,yield 语句返回 c 元素。

第三次 for 循环,提取 f 函数的 c 元素,输出 c,然后 f 中的循环就结束了,显示 fun End。随后在 for 循环中,f 也没有元素可以继续提取了,程序结束。

只要包含 yield 语句的函数都会返回一个 generator 的可循环对象，执行到 yield 语句时只返回一个值，等待调用循环提取。一旦调用循环提取后，函数又继续进行。

这个过程非常类似两个线程的协作过程，提供数据的一方与接收数据的一方协作进行。数据提供方使用 yield 语句提供数据，等待数据接收方取走数据，如果不取走，yield 语句就一直等待，一旦数据被取走，数据提供方就继续执行下一条 yield 语句，再次等待数据接收方取走数据。如此循环，一直到程序结束。

scrapy 框架使用的是异步执行的过程，因此大量使用 yield 语句。

## 4.3 使用 BeautifulSoup 爬取数据

4-3-A
知识讲解

4-3-B
操作演练

**任务目标**

使用 scrapy 创建一个爬虫程序，然后搭配使用 BeautifulSoup 解析并爬取模拟图书网站的数据。在本节中，我们主要学习 scrapy 与 BeautifulSoup 的混合使用方法。

### 4.3.1 创建模拟图书网站

#### 1. 创建网站模板

使用 books.csv 文件中的数据在 project4 的 templates 文件夹中创建一个模板文件 book.html，文件内容如下：

```html
<style>
.pic {display:inline-block;width:200px; vertical-align:top;margin:10px;}
.info { display:inline-block; width:600px;}
.liClass { list-style: none; margin:20px;}
.title { }
.price { margin: 10px;color:red; }
.attrs { margin: 10px;color: #666; }
h3 { display:inline-block;}}
.pl {color:#888;}
.author {}
.publisher {}
.date {}
.detail {}
.link {border: 1px solid ;}
a:link { color: blue; text-decoration: none; }
a:visited { color: blue; text-decoration: none; }
</style>
<div>
    <div class="pic">
     <img width="200"  src='/images/000002.jpg'>
  </div>
    <div class="info">
        <div class="title"><h3 style="display:inline-block"> Python 编程（第四版）</h3></div>
        <div class="author">
            <span class="pl">作者</span>:<span class="attrs">Mark</span>
```

```
            </div>
            <div class="publisher">
                <span class="pl">出版社</span>:<span class="attrs">中国电力出
版社</span>
            </div>
            <div class="date">
                <span class="pl">出版时间</span>:<span class="attrs">
2014-12-01</span>
            </div>
            <div class="price">
                <span class="pl">价格</span>:<span class="attrs">
¥136.60</span>
            </div>
            <div>简介:</div>
            <div class="detail">
                重磅推荐: Python 袖珍指南(第五版) 本书是 Python 应用的手册指南,它
涵盖了 Python 编程的方方面面,从系统管理到图形界面编程,从文本处理到网络编程,从数据库到语
言扩展。在这些主题的探讨中,作者提供了大量的实际代码,通过对这些代码的研读,可提高读者的
Python 编程水平以及实际业务问题的解决能力。
            </div>
        </div>
    </div>
```

### 2. 创建网站服务器程序

使用 Flask 设计服务器程序 server.py,如下:

```
import flask
app=flask.Flask(__name__,static_folder="images")
@app.route("/")
def index():
    return flask.render_template("book.html")
app.run()
```

运行这个服务器程序,使用浏览器访问"http://127.0.0.1:5000",就会看到图 4-3-1 所示的图书网站。

图 4-3-1 图书网站

## 4.3.2 解析网站的 HTML 代码

### 1. 获取网站的 HTML 代码

仍然使用 project4 下面的 scrapy 项目 demo,修改 project4\demo\demo\MySpider.py 程序,

在它的 parse()函数中获取 HTML 代码：

```
def parse(self, response):
    print(response.url)
    html=response.body.decode()
    soup=BeautifulSoup(html,"lxml")
    div=soup.find("div",attrs={"class":"info"})
```

其中，response.body 就是返回的网页二进制数据，html 就是 HTML 代码，用它创建一个 BeautifulSoup 对象 soup。

### 2. 爬取图书名称

图书名称包含在<div class="info">的<div class="title">的<h3>中，因此图书名称 bTitle 可以这样获取：

```
d=div.find("div",attrs={"class":"title"})
bTitle=d.find("h3").text.strip()
```

### 3. 爬取图书其他信息

作者信息包含在<div class="info">的<div class="author">的<span class="attrs">中，因此作者信息 bAuthor 可以这样获取：

```
d=div.find("div",attrs={"class":"author"})
bAuthor=d.find("span",attrs={"class":"attrs"}).text.strip()
```

同理，可以爬取图书出版社 bPublisher、出版时间 bDate、价格 bPrice、简介 bDetail 等信息，程序如下：

```
d = div.find("div", attrs={"class": "publisher"})
bPublisher = d.find("span", attrs={"class": "attrs"}).text.strip()
d = div.find("div", attrs={"class": "date"})
bDate = d.find("span", attrs={"class": "attrs"}).text.strip()
d = div.find("div", attrs={"class": "price"})
bPrice = d.find("span", attrs={"class": "attrs"}).text.strip()
bDetail = div.find("div", attrs={"class": "detail"}).text.strip()
```

## 4.3.3 爬取图书图像

### 1. 获取图像地址

图书图像包含在<div class="pic">的<img src=…>中，因此图像地址 src 可以这样获取：

```
src=soup.find("div",attrs={"class":"pic"}).find("img")["src"]
src=response.urljoin(src)
```

获取的 src 是相对地址，需要使用 response.urljoin()函数把它变成绝对地址，该函数使用当前的地址与获取的相对地址的组合将其变成绝对地址，如将 "/images/000002.jpg" 变成 http://127.0.0.1:5000/images/000002.jpg"。

### 2. 使用 scrapy 下载图像

获取图像地址 src 后可以应用 urllib.request.urlopen(src)请求这个 src，编写 download()函数进行下载。在前面的项目中，我们已经分析过这种下载方法可能会阻塞程序的进程，一个好的方法是使用一个子线程去执行 download()函数，即让下载变成一个异步的动作。

实际上，scrapy 本身就是一个异步执行的框架，在下载图像时不需要另外使用一个子线程去执行下载函数 download()，只要按 scrapy 的规则设计下载函数就能像子线程那样异步执行下载工作,而不会阻塞目前的程序。

使用 scrapy 下载图像时,首先使用图像地址 src 创建一个 scrapy.Request 对象,然后使用 yield 语句提交这个对象,在创建对象时规定了访问的地址是 src,访问完毕使用的回调函数是 download()函数,语句为:

```
yield scrapy.Request(url=src,callback=self.download)
```

这个过程类似 start_requests(self)函数中的 yield scrapy.Request(url=url,callback=self.parse)语句,因此 download()函数也类似 parse()函数,它有一个 response 对象,该对象就是访问 src 完成后的响应对象,而 response.body 就是网页的二进制数据,也就是图像数据。因此 download()函数如下:

```
def download(self,response):
    print(response.url)
    src = response.url.strip("/")
    p = src.rfind("/")
    fileName = src[p + 1:]
    f=open("download\\"+fileName,"wb")
    f.write(response.body)
    f.close()
    print("download ",fileName)
```

函数中的 response.url 就是图像地址 src,找到其最后一个"/"字符,分解出文件的名称 fileName,最后把图像数据 response.body 保存到 download 文件夹的 fileName 文件中。

### 4.3.4 编写爬虫程序

这里继续使用 project4\demo 的 scrapy 项目,根据前面的分析重新编写的 MySpider.py 程序如下:

```
import scrapy
from bs4 import BeautifulSoup
import os

class MySpider(scrapy.Spider):
    name = "mySpider"
    def start_requests(self):
        url ='http://127.0.0.1:5000'
        yield scrapy.Request(url=url,callback=self.parse)

    def parse(self, response):
        if not os.path.exists("download"):
            os.mkdir("download")
        print(response.url)
        html=response.body.decode()
        soup=BeautifulSoup(html,"lxml")
        div=soup.find("div",attrs={"class":"info"})
        d=div.find("div",attrs={"class":"title"})
        bTitle=d.find("h3").text.strip()
        d=div.find("div",attrs={"class":"author"})
        bAuthor=d.find("span",attrs={"class":"attrs"}).text.strip()
        d=div.find("div",attrs={"class":"publisher"})
        bPublisher=d.find("span",attrs={"class":"attrs"}).text.strip()
        d=div.find("div",attrs={"class":"date"})
```

```
            bDate=d.find("span",attrs={"class":"attrs"}).text.strip()
            d=div.find("div",attrs={"class":"price"})
            bPrice=d.find("span",attrs={"class":"attrs"}).text.strip()
            bDetail=div.find("div",attrs={"class":"detail"}).text.strip()
            print(bTitle)
            print(bAuthor)
            print(bPublisher)
            print(bDate)
            print(bPrice)
            print(bDetail)
            src=soup.find("div",attrs={"class":"pic"}).find("img")["src"]
            src=response.urljoin(src)
            yield scrapy.Request(url=src, callback=self.download)

    def download(self,response):
            print(response.url)
            src = response.url.strip("/")
            p = src.rfind("/")
            fileName = src[p + 1:]
            f=open("download\\"+fileName,"wb")
            f.write(response.body)
            f.close()
            print("download ",fileName)
```

## 4.3.5 执行爬虫程序

执行 demo 中的 run.py，可以看到爬虫程序成功爬取了网站的图书数据，得到如下结果。

```
http://127.0.0.1:5000
Python 编程（第四版）
Mark
中国电力出版社
2014-12-01
¥136.60
重磅推荐：Python 袖珍指南（第五版）本书是 Python 应用的手册指南，它涵盖了 Python 编程的方方面面，从系统管理到图形界面编程，从文本处理到网络编程，从数据库到语言扩展。在这些主题的探讨中，作者提供了大量的实际代码，通过对这些代码的研读，可提高读者的 Python 编程水平以及实际业务问题的解决能力。
http://127.0.0.1:5000/images/000002.jpg
download 000002.jpg
```

## 4.4 使用 XPath 查找元素

4-4-A    4-4-B    知识讲解    操作演练

 任务目标

使用 BeautifulSoup 能很好地查找 HTML 中的元素。scrapy 中也有强大的查找 HTML 元素的功能，那就是使用 XPath 方法。XPath 方法使用 XPath 语法，比 BeautifulSoup 的 find()、select() 函数要灵活，而且速度快。在本节中，我们主要学习使用 XPath 查找元素的方法。

### 4.4.1 scrapy 的 XPath 简介

scrapy 支持使用 XPath 查找元素，XPath 功能强大，能快速定位到要查找的元素。

例 4-4-1：使用 XPath 查找 HTML 中的元素。

```
from scrapy.selector import Selector
htmlText='''
<html><body>
<bookstore>
<book>
  <title lang="eng">Harry Potter</title>
  <price>29.99</price>
</book>
<book>
  <title lang="eng">Learning XML</title>
  <price>39.95</price>
</book>
</bookstore>
</body></html>
'''
selector=Selector(text=htmlText)
print(type(selector));
print(selector)
s=selector.xpath("//title")
print(type(s))
print(s)
```

结果：

```
class 'scrapy.selector.unified.Selector'>
<Selector xpath=None data='<html><body>\n<bookstore>\n<book>\n  <title'>
<class 'scrapy.selector.unified.SelectorList'>
[<Selector xpath='//title' data='<title lang="eng">Harry Potter</title>'>,
<Selector xpath='//title' data='<title lang="eng">Learning XML</title>'>]
```

下面分析程序的功能。

（1）语句：

```
from scrapy.selector import Selector
```

从 scrapy 中引入 Selector 类，这个类就是选择查找类。

（2）语句：

```
selector=Selector(text=htmlText)
```

使用 htmlText 的文字建立 Selector 类，即装载 HTML 文档，文档装载后就形成一个 Selector 对象，这样就可以使用 XPath 查找元素了。

（3）语句：

```
print(type(selector))
```

selector 是一个类型为 scrapy.selector.unified.Selector 的对象，这个类型是一个有 XPath 方法的类型。

（4）语句：

```
s=selector.xpath("//title")
```

这个方法用于在文档中查找所有的<title>元素，其中，"//"表示全文档搜索。

一般情况下，以下语句：

```
selector.xpath("//tagName")
```
表示在文档中搜索<tagName>元素,结果是一个 Selector 的列表。

(5)语句:
```
print(type(s))
```
由于有两个<title>元素,因此我们看到这是一个 scrapy.selector.unified.SelectorList 类,类似 scrapy.selector.unified.Selector 的列表。

(6)语句:
```
print(s)
```
我们看到 s 包含两个 Selector 对象,一个是<Selector xpath='//title' data='<title lang="eng">Harry Potter</title>'>,另一个是<Selector xpath='//title' data='<title lang="eng">Learning XML</title>'>。

一般 Selector 中用于搜索<tagName>的 HTML 元素的方法是:
```
selector.xpath("//tagName")
```
装载 HTML 文档后,selector=Selector(text=htmlText)得到的 selector 是对应全文档顶层的<html>元素,其中,"//"表示全文档搜索,结果是一个 Selector 的列表,哪怕只有一个元素,也是一个列表。例如:

selector.xpath("//body")搜索到一个<body>元素,结果是一个 Selector 的列表,包含一个 Selector 元素;

selector.xpath("//title")搜索到两个<title>元素,结果是一个 Selector 的列表,包含两个 Selector 元素;

selector.xpath("//book")搜索到两个<book>元素,结果是一个 Selector 的列表,包含两个 Selector 元素。

### 4.4.2 使用 XPath 查找 HTML 元素

(1)使用"//"表示文档下面的所有节点元素,使用"/"表示当前节点的下一级节点元素。

例 4-4-2:"//"与"/"的使用。

① 语句:
```
selector.xpath("//bookstore/book")
```
搜索<bookstore>下一级的<book>元素,找到两个。

② 语句:
```
selector.xpath("//body/book")
```
搜索<body>下一级的<book>元素,结果为空。

③ 语句:
```
selector.xpath("//body//book")
```
搜索<body>下的<book>元素,找到两个。

④ 语句:
```
selector.xpath("/body//book")
```
搜索文档下一级的<body>下的<book>元素,结果为空。这是因为文档下一级的元素是<html>元素,而不是<body>元素。

⑤ 语句：

```
selector.xpath("/html/body//book")
```

或者

```
selector.xpath("/html//book")
```

搜索<book>元素，找到两个。

⑥ 语句：

```
selector.xpath("//book/title")
```

搜索文档中所有<book>下一级的<title>元素，找到两个，结果与 selector.xpath("//title")、selector.xpath("//bookstore//title")的结果一样。

⑦ 语句：

```
selector.xpath("//book//price")
selector.xpath("//price")
```

两条语句的结果一致，都是找到两个<price>元素。

（2）使用 "." 表示当前节点元素，使用 XPath 可以连续调用，如果前一个 XPath 返回一个 Selector 的列表，那么这个列表可以连续调用 XPath，其功能是为每个列表元素调用 XPath，最后结果是全部元素调用 XPath 的汇总。

**例 4-4-3**：使用 "." 进行 XPath 连续调用。

```
from scrapy.selector import Selector
htmlText='''
<html>
<body>
<bookstore>
<title>books</title>
<book>
  <title>Novel</title>
  <title lang="eng">Harry Potter</title>
  <price>29.99</price>
</book>
<book>
  <title>TextBook</title>
  <title lang="eng">Learning XML</title>
  <price>39.95</price>
</book>
</bookstore>
</body></html>
'''
selector=Selector(text=htmlText)
s=selector.xpath("//book").xpath("./title")
for e in s:
       print(e)
```

结果：

```
<Selector xpath='//book/title' data='<title>Novel</title>'>
  <Selector xpath='//book/title' data='<title lang="eng">Harry Potter</title>'>
  <Selector xpath='//book/title' data='<title>TextBook</title>'>
  <Selector xpath='//book/title' data='<title lang="eng">Learning XML</title>'>
```

我们看到，selector.xpath("//book")首先搜索文档中所有的<book>元素，总共有两个，然后调用 xpath("./title")，从当前元素<book>开始往下一级搜索<title>元素，每个<book>元素中都有两个<title>元素，因此结果有 4 个<title>元素。

如果连续调用 XPath 时不指定是从前一个 XPath 的元素开始的，那么默认是从全文档开始的，结果会不一样。例如，使用语句

```
s=selector.xpath("//book").xpath("/title")
```

得到的结果是空的，因为后面的 xpath("/title")从文档的头部开始搜索<title>元素；而使用语句

```
s=selector.xpath("//book").xpath("//title")
```

得到的结果有 10 个元素，因为每个<book>元素都驱动 xpath("//title")在全文档搜索<title>元素，每次都搜索到 5 个<title>元素。

（3）如果 XPath 返回 Selector 对象列表，再次调用 extract()函数会得到这些对象元素的文本组成的列表，使用 extract_first()函数获取列表中第一个元素的文本值。如果列表为空，那么 extract_first()函数的值为 None。

而对于单一的 Selector 对象，调用 extract()函数就可以得到 Selector 对象对应的元素的文本值。单一的 Selector 对象没有 extract_first()函数。

例 4-4-4：extract()与 extract_first()函数的使用。

```
from scrapy.selector import Selector
htmlText='''
<html>
<body>
<bookstore>
<book id="b1">
  <title lang="english">Harry Potter</title>
  <price>29.99</price>
</book>
<book id="b2">
  <title lang="chinese">学习 XML</title>
  <price>39.95</price>
</book>
</bookstore>
</body></html>
'''
selector=Selector(text=htmlText)
s=selector.xpath("//book/price")
print(type(s),s)
s=selector.xpath("//book/price").extract()
print(type(s),s)
s=selector.xpath("//book/price").extract_first()
print(type(s),s)
```

结果：

```
<class 'scrapy.selector.unified.SelectorList'> [<Selector xpath='//book/price' data='<price>29.99</price>'>, <Selector xpath='//book/price' data='<price>39.95</price>'>]
<class 'list'> ['<price>29.99</price>', '<price>39.95</price>']
<class 'str'> <price>29.99</price>
```

由此可见，使用语句：

```
s=selector.xpath("//book/price")
```
得到的是 SelectorList 列表。

使用语句：
```
s=selector.xpath("//book/price").extract()
```
得到的是<price>元素的 Selector 对象对应的<price>元素的文本组成的列表，即：
```
['<price>29.99</price>', '<price>39.95</price>']
```
使用语句：
```
s=selector.xpath("//book/price").extract_first()
```
得到的是<price>元素的文本组成的列表的第一个元素的文本，即：
```
<price>29.99</price>
```

（4）XPath 使用 "/@attrName" 得到一个 Selector 元素的 attrName 属性节点对象，属性节点对象也是一个 Selector 对象，可以通过 extract()函数获取属性值。

例 4-4-5：获取元素属性值。
```
htmlText='''
<html>
<body>
<bookstore>
<book id="b1">
  <title lang="english">Harry Potter</title>
  <price>29.99</price>
</book>
<book id="b2">
  <title lang="chinese">学习 XML</title>
  <price>39.95</price>
</book>
</bookstore>
</body></html>
'''
selector=Selector(text=htmlText)
s=selector.xpath("//book/@id")
print(s)
print(s.extract())
for e in s:
    print(e.extract())
```

结果：
```
[<Selector xpath='//book/@id' data='b1'>, <Selector xpath='//book/@id' data='b2'>]
['b1', 'b2']
b1
b2
```

由此可见，使用语句：
```
s=selector.xpath("//book/@id")
```
得到的是两个<book>的 id 属性组成的 SelectorList 列表，即属性也是一个 Selector 对象；使用语句：
```
print(s.extract())
```
得到的是<book>的 id 属性的两个 Selector 对象的属性文本值组成的列表，即['b1', 'b2']。

在下面的语句中：
```
for e in s:
    print(e.extract())
```
每个 e 都是一个 Selector 对象，因此 extract()函数获取的是对象的属性值。

（5）XPath 使用"/text()"得到一个 Selector 元素包含的文本值节点对象，文本值节点对象也是一个 Selector 对象，可以通过 extract()函数获取文本值。

例 4-4-6：获取节点的文本值。

```
from scrapy.selector import Selector
htmlText='''
<html>
<body>
<bookstore>
<book id="b1">
  <title lang="english">Harry Potter</title>
  <price>29.99</price>
</book>
<book id="b2">
  <title lang="chinese">学习 XML</title>
  <price>39.95</price>
</book>
</bookstore>
</body></html>
'''
selector=Selector(text=htmlText)
s=selector.xpath("//book/title/text()")
print(s)
print(s.extract())
for e in s:
    print(e.extract())
```

结果：
```
[<Selector xpath='//book/title/text()' data='Harry Potter'>, <Selector xpath='//book/title/text()' data='学习 XML'>]
['Harry Potter', '学习 XML']
Harry Potter
学习 XML
```

由此可见，使用语句：
```
s=selector.xpath("//book/title/text()")
```
得到的也是 SelectorList 列表，即文本也是一个节点；使用语句：
```
print(s.extract())
```
得到的是文本节点的字符串值组成的列表，即['Harry Potter', '学习 XML']。

在下面的语句中：
```
for e in s:
    print(e.extract())
```
每个 e 都是一个 Selector 对象，因此 extract()函数获取的是对象的属性值。

值得注意的是，如果 element 的元素包含的文本不是单一的文本，那么可能会产生多个文本值。

例 4-4-7：多个文本值。

```
from scrapy.selector import Selector
htmlText='''
<html>
<body>
<bookstore>
<book id="b1">
  <title lang="english"><b>H</b>arry <b>P</b>otter</title>
  <price>29.99</price>
</book>
</bookstore>
</body></html>
'''
selector=Selector(text=htmlText)
s=selector.xpath("//book/title/text()")
print(s)
print(s.extract())
for e in s:
    print(e.extract())
```

结果：

```
[<Selector xpath='//book/title/text()' data='arry '>, <Selector xpath='//book/title/text()' data='otter'>]
['arry ', 'otter']
arry 
otter
```

由此可见，<title>中的文本值包含 arry 与 otter 两个字符串。

（6）XPath 使用 "tag[condition]" 来限定一个 tag 元素，其中 condition 是由这个 tag 的属性、文本等计算出的一个逻辑值。如果有多个条件，那么可以写成如下形式：

```
"tag[condition1][condition2]...[conditionN]"
```

或者：

```
"tag[condition1 and condition2 and ... and conditionN]"
```

例 4-4-8：使用 condition 限定 tag 元素。

```
from scrapy.selector import Selector
htmlText='''
<html>
<body>
<bookstore>
<book id="b1">
  <title lang="english">Harry Potter</title>
  <price>29.99</price>
</book>
<book id="b2">
  <title lang="chinese">学习 XML</title>
  <price>39.95</price>
</book>
</bookstore>
</body></html>
'''
selector=Selector(text=htmlText)
```

```
s=selector.xpath("//book/title[@lang='chinese']/text()")
print(s.extract_first())
s=selector.xpath("//book[@id='b1']/title")
print(s.extract_first())
```

结果:

```
学习 XML
<title lang="english">Harry Potter</title>
```

由此可见,语句

```
s=selector.xpath("//book/title[@lang='chinese']/text()")
```
用来搜索<book>下面属性为 lang='chinese'的<title>元素;语句

```
s=selector.xpath("//book[@id='b1']/title")
```
用来搜索属性为 id='b1'的<book>下面的<title>元素。

(7) XPath 可以使用 position()函数来确定一个元素的位置限制(元素的选择序号是从 1 开始编号的,而不是从 0 开始编号的),还可以使用 and、or 等构造复杂的表达式。

例 4-4-9:使用 position()函数来确定选择的元素。

```
from scrapy.selector import Selector
htmlText='''
<html>
<body>
<bookstore>
<book id="b1">
  <title lang="english">Harry Potter</title>
  <price>29.99</price>
</book>
<book id="b2">
  <title lang="chinese">学习 XML</title>
  <price>39.95</price>
</book>
</bookstore>
</body></html>
'''
selector=Selector(text=htmlText)
s=selector.xpath("//book[position()=1]/title")
print(s.extract_first())
s=selector.xpath("//book[position()=2]/title")
print(s.extract_first())
```

结果:

```
<title lang="english">Harry Potter</title>
<title lang="chinese">学习 XML</title>
```

其中,语句:

```
s=selector.xpath("//book[position()=1]/title")
s=selector.xpath("//book[position()=2]/title")
```
分别用于选择第一个、第二个<book>元素。

(8) XPath 使用 "*" 代表任何 element 节点,不包括 text、comment 的节点。

例4-4-10：使用"*"代表任何element节点。

```
from scrapy.selector import Selector
htmlText='''
<html>
<body>
<bookstore>
<book id="b1">
  <title lang="english">Harry Potter</title>
  <price>29.99</price>
</book>
<book id="b2">
  <title lang="chinese">学习 XML</title>
  <price>39.95</price>
</book>
</bookstore>
</body></html>
'''
selector=Selector(text=htmlText)
s=selector.xpath("//bookstore/*/title")
print(s.extract())
```

结果：

```
['<title lang="english">Harry Potter</title>', '<title lang="chinese">学习 XML</title>']
```

其中，语句：

```
s=selector.xpath("//bookstore/*/title")
```

用于搜索<bookstore>的孙子节点<title>，隔开中间一层的任何元素。

（9）XPath使用"@*"代表任何属性。

例4-4-11：使用"@*"代表任何属性。

```
from scrapy.selector import Selector
htmlText='''
<html>
<body>
<bookstore>
<book>
  <title lang="english">Harry Potter</title>
  <price>29.99</price>
</book>
<book id="b2">
  <title lang="chinese">学习 XML</title>
  <price>39.95</price>
</book>
</bookstore>
</body></html>
'''
selector=Selector(text=htmlText)
s=selector.xpath("//book[@*]/title")
print(s.extract())
s=selector.xpath("//@*")
print(s.extract())
```

结果:

```
['<title lang="chinese">学习 XML</title>']
['english', 'b2', 'chinese']
```

其中,语句:

```
s=selector.xpath("//book[@*]/title")
```

用于搜索任何包含属性的<book>元素下面的<title>,结果搜索到第二个<book>;语句:

```
s=selector.xpath("//@*")
```

用于搜索文档中的所有属性节点。

(10) XPath 使用"element/parent::*"选择 element 的父节点,这个节点只有一个。如果写成"element/parent::tag",就选择 element 的 tag 父节点,除非 element 的父节点正好为<tag>节点,不然就为 None。

例 4-4-12:XPath 搜索元素的父节点。

```
from scrapy.selector import Selector
htmlText='''
<html>
<body>
<bookstore>
<book>
  <title lang="english">Harry Potter</title>
  <price>29.99</price>
</book>
<book id="b2">
  <title lang="chinese">学习 XML</title>
  <price>39.95</price>
</book>
</bookstore>
</body></html>
'''
selector=Selector(text=htmlText)
s=selector.xpath("//title[@lang='chinese']/parent::*")
print(s.extract())
```

结果:

```
['<book id="b2">\n  <title lang="chinese">学习 XML</title>\n  <price>39.95</price>\n</book>']
```

其中,语句:

```
s=selector.xpath("//title[@lang='chinese']/parent::*")
```

用于查找属性为 lang='chinese'的<title>元素的父节点,即 id='b2'的<book>元素节点。

(11) XPath 使用"element/following-sibling::*"搜索 element 后面的同级的所有兄弟节点,使用"element/following-sibling::*[position()=1]"搜索 element 后面的同级的第一个兄弟节点。

例 4-4-13:搜索后面的兄弟节点。

```
from scrapy.selector import Selector
htmlText="<a>A1</a><b>B1</b><c>C1</c><d>D<e>E</e></d><b>B2</b><c>C2</c>"
selector=Selector(text=htmlText)
s=selector.xpath("//a/following-sibling::*")
print(s.extract())
s=selector.xpath("//a/following-sibling::*[position()=1]")
```

```
print(s.extract())
s=selector.xpath("//b[position()=1]/following-sibling::*")
print(s.extract())
s=selector.xpath("//b[position()=1]/following-sibling::*[position()=1]")
print(s.extract())
```

结果:

```
['<b>B1</b>', '<c>C1</c>', '<d>D<e>E</e></d>', '<b>B2</b>', '<c>C2</c>']
['<b>B1</b>']
['<c>C1</c>', '<d>D<e>E</e></d>', '<b>B2</b>', '<c>C2</c>']
['<c>C1</c>']
```

其中,语句:

```
s=selector.xpath("//b[position()=1]/following-sibling::*[position()=1]")
```

用于搜索第一个<b>节点后面的第一个兄弟节点,即<c>C1</c>节点。

(12) XPath 使用 "element/preceding-sibling::*" 搜索 element 前面的同级的所有兄弟节点,使用 "element/preceding-sibling::*[position()=1]" 搜索 element 前面的同级的第一个兄弟节点。

**例 4-4-14**: 搜索前面的兄弟节点。

```
from scrapy.selector import Selector
htmlText="<a>A1</a><b>B1</b><c>C1</c><d>D<e>E</e></d><b>B2</b><c>C2</c>"
selector=Selector(text=htmlText)
s=selector.xpath("//a/preceding-sibling::*")
print(s.extract())
s=selector.xpath("//b/preceding-sibling::*[position()=1]")
print(s.extract())
s=selector.xpath("//b[position()=2]/preceding-sibling::*")
print(s.extract())
s=selector.xpath("//b[position()=2]/preceding-sibling::*[position()=1]")
print(s.extract())
```

结果:

```
[]
['<a>A1</a>', '<d>D<e>E</e></d>']
['<a>A1</a>', '<b>B1</b>', '<c>C1</c>', '<d>D<e>E</e></d>']
['<d>D<e>E</e></d>']
```

其中,语句:

```
s=selector.xpath("//b/preceding-sibling::*[position()=1]")
```

用于搜索所有<b>前面的第一个兄弟节点,因为有两个<b>节点,所以结果是['<a>A1</a>', '<d>D<e>E</e></d>']。

### 4.4.3 使用 XPath 与 BeautifulSoup

在解析 HTML 的数据时,XPath 与 BeautifulSoup 各有千秋,读者可以任意选择一种方法,但是在解析同一段代码时两者有时候会有些不同,读者在使用的过程中应当注意。

**例 4-4-15**: 解析空字符串的区别。

```
from scrapy.selector import Selector
from bs4 import BeautifulSoup
html="<div>Testing<span></span></div>"
soup=BeautifulSoup(html,"lxml")
s=soup.find("span").text
print(s,len(s))
```

```
selector=Selector(text=html)
s=selector.xpath("//span/text()").extract_first()
print(s)
```

结果：

```
0
None
```

上述结果说明，BeautifulSoup 解析<span></span>时得到的文本是一个空字符串，而 XPath 得到的是 None 值。

## 4.5 爬取关联网页数据

4-5-A　　4-5-B

知识讲解　　操作演练

### 任务目标

使用已有数据创建一个图书网站，这个网站包含多个网页，网页之间通过超链接相互关联，使用 scrapy 编写一个爬虫程序爬取每个网页的数据。使用 scrapy 爬取各个网页的顺序与使用深度优先法及广度优先法爬取各个网页的顺序不同。当使用深度优先法和广度优先法时，爬取各个网页的顺序是确定的，但是使用 scrapy 爬取各个网页的顺序是不确定的，是随机的，这就是 scrapy 的分布式特性。

### 4.5.1　创建模拟图书网站

**1. 创建网站模板**

在 project4\templates 中创建几个网页文件，分别是 books.html、database.html、program.html、network.html、mysql.html、python.html 及 java.html，这些网页互相关联，每个文件都有一个由<h3>定义的标题。

（1）books.html：

```
<h3>计算机</h3>
<ul>
<li><a href="database.html">数据库</a></li>
<li><a href="program.html">程序设计</a></li>
<li><a href="network.html">计算机网络</a></li>
</ul>
```

（2）database.html：

```
<h3>数据库</h3>
<ul>
<li><a href="mysql.htm">MySQL 数据库</a></li>
</ul>
<a href="books.html">Home</a>
```

（3）program.html：

```
<h3>程序设计</h3>
<ul>
<li><a href="python.html">Python 程序设计</a></li>
<li><a href="java.html">Java 程序设计</a></li>
</ul>
<a href="books.html">Home</a>
```

（4）network.html：

```
<h3>计算机网络</h3>
<a href="books.html">Home</a>
```

（5）mysql.html：

```
<h3>MySQL 数据库</h3>
<a href="books.html">Home</a>
```

（6）python.html：

```
<h3>Python 程序设计</h3>
<a href="books.html">Home</a>
```

（7）java.html：

```
<h3>Java 程序设计</h3>
<a href="books.html">Home</a>
```

2. 创建服务器程序

网站以 books.html 为入口，为了让这个网站更好地模拟真实的网站，我们在程序中提交每一个网页时均使用了随机时间延迟，time.sleep(random.randint(0,2))语句会把一个网页的响应时间延迟 0～2s，因此设计的服务器程序 server.py 如下：

```python
import flask
import time
import random

app=flask.Flask(__name__,static_folder="images")
@app.route("/")
def index():
    return flask.render_template("books.html")
@app.route("/<name>")
def show(name):
    time.sleep(random.randint(0,2))
    if name.strip()=="":
        name="books.html"
    return flask.render_template(name)
app.run()
```

运行服务器程序，使用浏览器访问 "http://127.0.0.1:5000"，就会看到图 4-5-1 所示的网页，单击超链接后会跳转到数据库、程序设计、计算机网络等网页。这些网页的结构实际上是一棵树，如图 4-5-2 所示。

图 4-5-1　关联网页　　　　　　　　图 4-5-2　Web 网站结构

## 4.5.2 程序爬取网页的顺序

### 1. 深度优先爬取网页的顺序

深度优先法也就是递归法，采用此方法编写爬取程序 spider.py，如下：

```python
import scrapy
import urllib.request

def spider(url):
    global urls
    try:
        urls.append(url)
        resp=urllib.request.urlopen(url)
        html=resp.read().decode()
        selector = scrapy.Selector(text=html)
        print(selector.xpath("//h3/text()").extract_first())
        links = selector.xpath("//a/@href").extract()
        for link in links:
            url = urllib.request.urljoin(url,link)
            if not url in urls:
                spider(url)
    except Exception as err:
        print(err)
urls=[]
spider("http://127.0.0.1:5000/books.html")
```

执行该程序，结果如下：

```
计算机
数据库
MySQL 数据库
程序设计
Python 程序设计
Java 程序设计
计算机网络
```

### 2. 广度优先爬取网页的顺序

采用广度优先法爬取网站要使用一个队列来进行，爬取程序 spider.py 如下：

```python
import scrapy
import urllib.request

class Queue:
    def __init__(self):
        self.st=[]
    def fetch(self):
        return self.st.pop(0)
    def enter(self,obj):
        self.st.append(obj)
    def empty(self):
        return len(self.st)==0

def spider(url):
```

```
        try:
            urls=[]
            queue=Queue()
            queue.enter(url)
            urls.append(url)
            while not queue.empty():
                url=queue.fetch()
                resp=urllib.request.urlopen(url)
                html=resp.read().decode()
                selector = scrapy.Selector(text=html)
                print(selector.xpath("//h3/text()").extract_first())
                links = selector.xpath("//a/@href").extract()
                for link in links:
                    url = urllib.request.urljoin(url,link)
                    if not url in urls:
                        queue.enter(url)
        except Exception as err:
            print(err)

spider("http://127.0.0.1:5000/books.html")
```

执行该程序，结果如下：

```
计算机
数据库
程序设计
计算机网络
MySQL 数据库
Python 程序设计
Java 程序设计
```

### 3. scrapy 爬取网页的顺序

现在使用 scrapy 编写一个爬虫程序以爬取这个网站所有网页的<h3>的文本，并查看 scrapy 是如何遍历各个网页的。

仍然使用相同的 scrapy 项目，重新编写 MySpider.py 程序，如下：

```
import scrapy
class MySpider(scrapy.Spider):
    name = "mySpider"
    def start_requests(self):
        url='http://127.0.0.1:5000/books.html'
        yield scrapy.Request(url=url, callback=self.parse)

    def parse(self, response):
        try:
            data=response.body.decode()
            selector=scrapy.Selector(text=data)
            print(selector.xpath("//h3/text()").extract_first())
            links=selector.xpath("//a/@href").extract()
            for link in links:
                url=response.urljoin(link)
                yield scrapy.Request(url=url,callback=self.parse)
```

```
        except Exception as err:
              print(err)
```

执行该程序，结果如下：

计算机
数据库
MySQL 数据库
程序设计
Python 程序设计
计算机网络
Java 程序设计

下面分析程序的执行过程。

（1）语句：

```
'http://127.0.0.1:5000/books.html'
```

这是入口地址，访问这个地址成功后会回调 parse()函数。

（2）语句：

```
def parse(self, response):
```

这是回调函数，该函数的 response 对象包含网站返回的信息。

（3）语句：

```
data=response.body.decode()
selector=scrapy.Selector(text=data)
```

网站返回的 response.body 的二进制数据通过 decode()函数转换为文本，然后建立 Selector 对象。

（4）语句：

```
print(selector.xpath("//h3/text()").extract_first())
```

获取网页中<h3>的文本，这就是要爬取的数据。简单起见，这个数据只有一项。

（5）语句：

```
links=selector.xpath("//a/@href").extract()
```

获取所有<a href=...>的 href 值，组成 links 列表。

（6）语句：

```
for link in links:
    url=response.urljoin(link)
    yield scrapy.Request(url=url,callback=self.parse)
```

访问 links 的每个 link，通过 urljoin()函数与 response.url 地址组合成完整的 url。再次建立 Request 对象，回调函数仍然为 parse()，即这个 parse()函数会被递归调用。其中使用 yield 语句返回每个 Request 对象，这是 scrapy 程序的要求。

很显然，网址"http://127.0.0.1:5000/books.html"是会被反复获取的，但是 scrapy 只访问它一次，即 scrapy 会自动记录已经访问过的网址，下次遇到相同的网址就不再访问了。

### 4.5.3 理解 scrapy 分布式

无论是采用深度优先法还是广度优先法爬取网站的网页，爬取时都是一个网页接着一个网页进行的，如果一个网页的响应速度比较慢，那么会阻塞程序对后面网页的爬取。

从 scrapy 程序的执行结果可以看到，scrapy 访问网页既不是按深度优先的顺序进行的，也不是按广度优先的顺序进行的。因为在 yield 提交 scrapy.Request(url=url,callback=self.parse) 请求后，这些请求会被 scrapy 管理，并异步访问每个网页，哪个网页响应快就先得到这个网页的结果，所以访问顺序是不确定的，这正是 scrapy 的分布式的特性。在这种分布式系统中，不会因为一个网页响应慢而阻塞程序对后面网页的爬取。

## 4.6 使用 XPath 爬取数据

**任务目标**

使用 scrapy 编写一个爬虫程序，使用 XPath 解析，爬取模拟图书网站的数据，将爬取的数据存储到数据库。在本节中，我们主要学习在 scrapy 中使用 XPath 的方法。

### 4.6.1 创建模拟图书网站

**1. 创建网站模板**

使用 books.csv 文件中的数据在 project4 的 templates 文件夹中创建一个模板文件 book.html，内容如下：

```
<style>
.pic {display:inline-block;width:200px; vertical-align:top;margin:10px;}
.info { display:inline-block; width:800px;}
.liClass { list-style: none; margin:20px;}
.title { }
.price { margin: 10px;color:red; }
.attrs { margin: 10px;color: #666; }
h3 { display:inline-block;}}
.pl {color:#888;}
.author {}
.publisher {}
.date {}
.detail {}
.link {border: 1px solid ;}
a:link { color: blue; text-decoration: none; }
a:visited { color: blue; text-decoration: none; }
</style>
<div>
<ul>
<li class="liClass"  >
    <div class="pic">
      <img width="200"  src='/images/000001.jpg'>
   </div>
    <div class="info">
        <div class="title"><h3 style="display:inline-block"> Python 编程 从入门到实践</h3></div>
        <div class="author">
            <span class="pl">作者</span>:<span class="attrs">埃里克·马瑟斯
```

```html
</span>
            </div>
            <div class="publisher">
                <span class="pl">出版社</span>:<span  class="attrs">人民邮电出版社</span>
            </div>
            <div class="date">
                <span class="pl">出版时间</span>:<span class="attrs">2016-07-01</span>
            </div>
            <div class="price">
                <span class="pl">价格</span>:<span class="attrs">¥62.00</span>
            </div>
                <div>简介:</div>
            <div class="detail">
                    上到有编程基础的程序员，下到10岁少年，想入门Python并达到可以开发实际项目的水平。本书是一本全面的从入门到实践的Python编程教程，带领读者快速掌握编程基础知识、编写出能解决实际问题的代码并开发复杂项目。 书中内容分为基础篇和实战篇两部分。基础篇介绍基本的编程概念，如列表、字典、类和循环，并指导读者编写整洁且易于理解的代码。另外还介绍如何让程序能够与用户交互，以及如何在代码运行前进行测试。实战篇介绍如何利用新学到的知识开发功能丰富的项目：2D游戏《外星人入侵》，数据可视化实战，Web应用程序。
            </div>
        </div>
    </li>
        <div></div>

    <li class="liClass"  >
       <div class="pic">
          <img width="200"   src='/images/000002.jpg'>
       </div>
       <div class="info">
            <div class="title"><h3 style="display:inline-block"> Python编程（第四版）</h3></div>
            <div class="author">
                <span class="pl">作者</span>:<span class="attrs">Mark</span>
            </div>
            <div class="publisher">
                <span class="pl">出版社</span>:<span   class="attrs">中国电力出版社</span>
            </div>
            <div class="date">
                <span class="pl">出版时间</span>:<span class="attrs">2014-12-01</span>
            </div>
            <div class="price">
                <span class="pl">价格</span>:<span class="attrs">¥136.60</span>
            </div>
```

```
            <div>简介:</div>
            <div class="detail">
                    重磅推荐: Python 袖珍指南(第五版) 本书是 Python 应用的手册指南,它
涵盖了 Python 编程的方方面面,从系统管理到图形界面编程,从文本处理到网络编程,从数据库到语
言扩展。在这些主题的探讨中,作者提供了大量的实际代码,通过对这些代码的研读,可提高读者的
Python 编程水平以及实际业务问题的解决能力。
            </div>
        </div>
    </li>
</ul>
</div>
```

### 2. 创建网站服务器程序

使用 Flask 设计的服务器程序 server.py 如下:

```
import flask
app=flask.Flask(__name__,static_folder="images")
@app.route("/")
def index():
    return flask.render_template("book.html")
app.run()
```

运行这个服务器程序,使用浏览器访问"http://127.0.0.1:5000",就会看到图 4-6-1 所示的图书网站。

图 4-6-1 图书网站

### 4.6.2 解析网站的 HTML 代码

#### 1. 查找<li>对象

仍然使用 project4 下面的 scrapy 项目 demo,修改 project4\demo\demo\MySpider.py 程序,在它的 parse()函数中获取 HTML 代码并创建 Selector 对象:

```
    def parse(self, response):
        print(response.url)
        selector=Selector(text=response.body.decode())
        lis=selector.xpath("//ul/li")
```

```
    for li in lis:
            div=li.xpath(".//div[@class='info']")
            #......
```

然后获取<ul>下面所有的<li>元素，循环每个<li>对象，图书的数据就存储在<li>的<div class="info">中。

### 2. 爬取图书名称

图书名称包含在<div class="info">的<div class="title">的<h3>中，因此图书名称 bTitle 可以这样获取：

```
d=div.xpath(".//div[@class='title']")
bTitle=d.xpath("./h3/text()").extract_first().strip()
```

### 3. 爬取图书其他信息

作者信息包含在<div class="info">的<div class="author">的<span class="attrs">中，因此作者 bAuthor 信息可以这样获取：

```
bAuthor = d.xpath("./span[@class='attrs']/text()").extract_first().strip()
d = div.xpath(".//div[@class='publisher']")
```

同理，可以爬取图书出版社 bPublisher、出版时间 bDate、价格 bPrice、简介 bDetail 等信息，程序如下：

```
bPublisher = d.xpath("./span[@class='attrs']/text()").extract_first().strip()
d = div.xpath(".//div[@class='date']")
bDate=d.xpath("./span[@class='attrs']/text()").extract_first().strip()
d=div.xpath(".//div[@class='price']")
bPrice = d.xpath("./span[@class='attrs']/text()").extract_first().strip()
bDetail = div.xpath(".//div[@class='detail']/text()").extract_first().strip()
```

## 4.6.3 爬取图书图像

### 1. 获取图像地址与名称

图书图像包含在<div class="pic">的<img src=...>中，因此图像地址 src 可以这样获取：

```
src=li.xpath(".//div[@class='pic']//img/@src").extract_first()
src=response.urljoin(src).strip("/")
```

获取的 src 是相对地址，需要使用 response.urljoin()函数把它变成绝对地址，该函数使用当前的地址与获取的相对地址的组合将其变成绝对地址。为了把图像名称与图书的其他数据一起存储，我们通过 parse()函数进一步得到图像名称 bImage：

```
p=src.rfind("/")
bImage=src[p+1:]
```

### 2. 传递图像名称到下载函数

实际上，scrapy.Request 中有一个字典 meta，在创建 scrapy.Request 时可以把图像名称存储到这个字典中，然后使用 yield 语句提交这个对象，例如：

```
request=scrapy.Request(url=src,callback=self.download)
request.meta["bImage"]=bImage
yield request
```

访问 src 结束后回调 download()函数，字典 meta 会被自动传递给 download()中的

response 对象,于是就可以从 response.meta["bImage"]中取出图像名称,即:

```
def download(self, response):
    bImage = response.meta["bImage"]
    f = open("download\\" + bImage, "wb")
    f.write(response.body)
    f.close()
```

这个 meta 字典很有用,它是请求对象 scrapy.Request 与响应对象 response 传递数据的重要途径。

### 4.6.4 设计数据库存储

设计一个 SQLite3 数据库 books.db 来存储下载的数据,数据库中有一张 books 表,其结构如表 4-6-1 所示。

表 4-6-1 books 表结构

| 字段名称 | 类型 | 说明 |
| --- | --- | --- |
| ID | varchar(8) | 编号(关键字) |
| bTitle | varchar(256) | 图书名称 |
| bAuthor | varchar(256) | 作者 |
| bPublisher | varchar(256) | 出版社 |
| bDate | varchar(256) | 出版时间 |
| bPrice | varchar(256) | 价格 |
| bDetail | text | 简介 |
| bImage | varchar(256) | 图像名称 |

### 4.6.5 编写爬虫程序

继续使用 project4\demo 的 scrapy 项目,根据前面的分析重新编写 MySpider.py 程序,如下:

```
from scrapy.selector import Selector
import sqlite3
import os

class MySpider(scrapy.Spider):
    name = "mySpider"
    def start_requests(self):
        url ='http://127.0.0.1:5000'
        yield scrapy.Request(url=url,callback=self.parse)

    def openDB(self):
        self.con=sqlite3.connect("books.db")
        self.cursor=self.con.cursor()
        try:
            self.cursor.execute("drop table books")
```

```
                except:
                    pass
                sql="create table books (ID varchar(8) primary key,bTitle
varchar(256),bAuthor varchar(256),bPublisher varchar(256),bDate
varchar(256),bPrice varchar(256),bDetail text,bImage varchar(256))"
                self.cursor.execute(sql)
                self.count=0
                if not os.path.exists("download"):
                    os.mkdir("download")

        def closeDB(self):
            self.con.commit()
            self.con.close()

        def insertDB(self,ID,bTitle,bAuthor,bPublisher,bDate,bPrice,
bDetail,bImage):
            try:
                sql="insert into books (ID,bTitle,bAuthor,bPublisher,
bDate,bPrice,bDetail,bImage) values (?,?,?,?,?,?,?,?)"
                self.cursor.execute(sql,[ID,bTitle,bAuthor,bPublisher,
bDate,bPrice,bDetail,bImage])
            except Exception as err:
                print(err)

        def showDB(self):
            self.cursor.execute("select ID,bTitle,bAuthor,bPublisher,
bDate,bPrice,bDetail,bImage from books")
            rows=self.cursor.fetchall()
            for row in rows:
                for r in row:
                    print(r)

        def parse(self, response):
            self.openDB()
            selector=Selector(text=response.body.decode())
            lis=selector.xpath("//ul/li")
            for li in lis:
                div=li.xpath(".//div[@class='info']")
                d=div.xpath(".//div[@class='title']")
                bTitle=d.xpath("./h3/text()").extract_first().strip()
                d=div.xpath(".//div[@class='author']")
                bAuthor=d.xpath("./span[@class='attrs']/text()").
extract_first().strip()
                d=div.xpath(".//div[@class='publisher']")
                bPublisher=d.xpath("./span[@class='attrs']/text()").
extract_first().strip()
                d=div.xpath(".//div[@class='date']")
                bDate=d.xpath("./span[@class='attrs']/text()").extract_
first().strip()
                d=div.xpath(".//div[@class='price']")
                bPrice=d.xpath("./span[@class='attrs']/text()").extract_
first().strip()
                bDetail=div.xpath(".//div[@class='detail']/text()").
extract_first().strip()
```

```
                src=li.xpath(".//div[@class='pic']//img/@src").extract_
first()
                src=response.urljoin(src).strip("/")
                p=src.rfind("/")
                bImage=src[p+1:]
                request=scrapy.Request(url=src,callback=self.download)
                request.meta["bImage"]=bImage
                yield request
                self.count+=1
                ID="%06d" % self.count
                self.insertDB(ID,bTitle,bAuthor,bPublisher,bDate,bPrice,
bDetail,bImage)
            self.showDB()
            self.closeDB()

        def download(self,response):
            bImage = response.meta["bImage"]
            f=open("download\\"+bImage,"wb")
            f.write(response.body)
            f.close()
```

程序中使用 self.count 记录图书的数量，使用长度为 6 位的字符串 ID 作为关键字，调用 insertDB() 函数把记录存储到 books 表中。

### 4.6.6 执行爬虫程序

执行爬虫程序，爬取网站的图书数据与图像，将爬取到的图书数据和图像存储到数据库 books.db 中，并将下载的图像存储到 download 文件夹中，结果如下：

```
000001
Python 编程 从入门到实践
埃里克·马瑟斯
人民邮电出版社
2016-07-01
¥62.00
```

上到有编程基础的程序员，下到 10 岁少年，想入门 Python 并达到可以开发实际项目的水平。本书是一本全面的从入门到实践的 Python 编程教程，带领读者快速掌握编程基础知识、编写出能解决实际问题的代码并开发复杂项目。书中内容分为基础篇和实战篇两部分。基础篇介绍基本的编程概念，如列表、字典、类和循环，并指导读者编写整洁且易于理解的代码。另外还介绍如何让程序能够与用户交互，以及如何在代码运行前进行测试。实战篇介绍如何利用新学到的知识开发功能丰富的项目：2D 游戏《外星人入侵》，数据可视化实战，Web 应用程序。

```
000001.jpg
000002
Python 编程（第四版）
Mark
中国电力出版社
2014-12-01
¥136.60
```

重磅推荐：Python 袖珍指南（第五版）本书是 Python 应用的手册指南，它涵盖了 Python 编程的方方面面，从系统管理到图形界面编程，从文本处理到网络编程，从数据库到语言扩展。在这些主题的探讨中，作者提供了大量的实际代码，通过对这些代码的研读，可提高读者的 Python 编程水平以及实际业务问题的解决能力。

000002.jpg

## 4.7 使用管道存储数据

4-7-A　4-7-B

知识讲解　操作演练

**任务目标**

从一个网站爬取到数据后往往要将数据存储到数据库中，scrapy 框架中有十分方便的存储方法，即管道方法。scrapy 把每次爬取的数据推送到管道中，这个管道连通一个存储类对象，这个存储类对象再把数据存储到数据库中。在本节中，我们主要学习这种管道技术。

### 4.7.1 创建模拟图书网站

**1. 创建网站模板**

在 project4\templates 中创建一个模板文件 book.html，这个模板文件包含 bImage、bTitle、bAuthor、bPublisher、bDate、bPrice 与 bDetails 等参数，并使用 books 列表参数构造一个循环，用来显示多本图书，它的内容如下：

```
<style>
.pic {display:inline-block;width:200px; vertical-align:top;margin:10px;}
.info { display:inline-block; width:800px;}
.liClass { list-style: none; margin:20px;}
.title { }
.price { margin: 10px;color:red; }
.attrs { margin: 10px;color: #666; }
h3 { display:inline-block;}}
.pl {color:#888;}
.author {}
.publisher {}
.date {}
.detail {}
.link {border: 1px solid ;}
a:link { color: blue; text-decoration: none; }
a:visited { color: blue; text-decoration: none; }
</style>
<div>
<ul>
{% for b in books %}
<li class="liClass"  >
    <div class="pic">
      <img width="200"  src='/images/{{b["bImage"]}}'>
   </div>
    <div class="info">
       <div class="title"><h3 style="display:inline-block">{{b["bTitle"]}}</h3></div>
       <div class="author">
          <span class="pl">作者</span>:<span class="attrs">{{b["bAuthor"]}}</span>
       </div>
```

```html
            <div class="publisher">
                <span class="pl">出版社</span>:<span class="attrs">{{b["bPublisher"]}}</span>
            </div>
            <div class="date">
                <span class="pl">出版时间</span>:<span class="attrs">{{b["bDate"]}}</span>
            </div>
            <div class="price">
                <span class="pl">价格</span>:<span class="attrs">{{b["bPrice"]}}</span>
            </div>
                <div>简介:</div>
             <div class="detail">
                 {{b["bDetail"]}}
             </div>
      </div>
  </li>
        <div></div>
  {% endfor %}
  </ul>
  </div>
```

### 2. 创建网站服务器程序

读取 books.csv 文件中的 10 本图书信息,并将信息传递到 book.html 模板文件,形成一个网页。服务器程序 server.py 如下:

```python
import flask
app=flask.Flask(__name__,static_folder="images")

@app.route("/")
def show():
    books=[]
    try:
        fobj=open("books.csv","r",encoding="utf-8")
        rows=fobj.readlines()
        for i in range(1,11):
            row=rows[i].strip("\n")
            if row!="":
                s=row.split(",")
                m={}
                m["ID"] = s[0]
                m["bImage"] = s[0] + s[7]
                m["bTitle"] = s[1]
                m["bAuthor"] = s[2]
                m["bPublisher"] = s[3]
                m["bDate"] = s[4]
                m["bPrice"] = s[5]
                m["bDetail"] = s[6]
                books.append(m)
        fobj.close()
    except Exception as err:
```

```
        print(err)
    return flask.render_template("book.html",books=books)
app.run()
```

运行这个服务器程序，使用浏览器访问"http://127.0.0.1:5000"，就会看到图 4-6-1 所示的图书网站。

## 4.7.2 编写数据字段类

在 4.5 节中使用 MySpider 类中的 parse()函数爬取到了图书数据，并调用 insertDB()函数完成了数据的存储。其中，调用 insertDB()函数是一个同步的过程，即 insertDB()会阻塞 parse()函数的执行，这一点与 scrapy 的分布式异步执行规则有些相悖。

实际上，scrapy 建议尽量异步存储数据，这就是 scrapy 的数据管道存储机制。具体来说就是首先使用 demo\demo\items.py 文件建立一个数据类 BookItem，然后在 parse()函数爬取到数据时创建一个数据类 BookItem 对象，最后使用 yield 语句把数据对象推送到 demo\dcmo\pipelines.py 文件的数据处理类 BookPipeline 进行数据存储。

由于 yield 语句提交的数据由 scrapy 统一协调，这个过程是异步的，因此数据存储是异步的。在 demo\demo 文件夹中有一个 items.py 文件，打开这个文件，根据爬取数据字段建立一个数据字段类 BookItem，如下：

```
import scrapy
class BookItem(scrapy.Item):
    # define the fields for your item here like:
    bTitle = scrapy.Field()
    bAuthor = scrapy.Field()
    bPublisher = scrapy.Field()
    bDate = scrapy.Field()
    bPrice = scrapy.Field()
    bDetail = scrapy.Field()
    bImage = scrapy.Field()
```

这个类命名为 BookItem，它必须从 scrapy.Item 派生而来，其中定义了 bTitle、bAuthor、bPublisher、bDate、bPrice、bDetail、bImage 等数据字段，每个字段都由 scrapy.Field()确定。scrapy.Field()是 scrapy 规定的一个字段对象，字段可以是任意数据类型。

如果 item 是一个 BookItem 对象，那么可以通过 item["字段名称"]来设置或者获取该字段的值，例如：

```
item=BookItem()
item["bTitle"]="Python 程序设计"
item["bAuthor"]="James"
item["bPublisher"]="清华大学出版社"
print(item["bTitle"])
print(item["bAuthor"])
print(item["bPublisher"])
```

结果：
```
Python 程序设计
James
清华大学出版社
```

### 4.7.3 编写爬虫程序类

使用 scrapy 的管道机制存储数据，demo\demo\spiders 中的 MySpider 类就不再负责数据的存储工作，它的任务是爬取数据并把数据组织在一个 BookItem 对象 item 中，然后使用 yield 语句提交这个 item 对象即可，因此 MySpider.py 修改如下：

```python
import scrapy
from scrapy.selector import Selector
from demo.items import BookItem

class MySpider(scrapy.Spider):
    name = "mySpider"
    def start_requests(self):
        url ='http://127.0.0.1:5000'
        yield scrapy.Request(url=url,callback=self.parse)

    def parse(self, response):
        selector=Selector(text=response.body.decode())
        lis=selector.xpath("//ul/li")
        for li in lis:
            div=li.xpath(".//div[@class='info']")
            d=div.xpath(".//div[@class='title']")
            bTitle=d.xpath("./h3/text()").extract_first().strip()
            d=div.xpath(".//div[@class='author']")
            bAuthor=d.xpath("./span[@class='attrs']/text()").extract_first().strip()
            d=div.xpath(".//div[@class='publisher']")
            bPublisher=d.xpath("./span[@class='attrs']/text()").extract_first().strip()
            d=div.xpath(".//div[@class='date']")
            bDate=d.xpath("./span[@class='attrs']/text()").extract_first().strip()
            d=div.xpath(".//div[@class='price']")
            bPrice=d.xpath("./span[@class='attrs']/text()").extract_first().strip()
            bDetail=div.xpath(".//div[@class='detail']/text()").extract_first().strip()
            src=li.xpath(".//div[@class='pic']//img/@src").extract_first()
            src=response.urljoin(src).strip("/")
            p=src.rfind("/")
            bImage=src[p+1:]
            request=scrapy.Request(url=src,callback=self.download)
            request.meta["bImage"]=bImage
            yield request
            #创建BookItem对象
            item=BookItem()
            item["bTitle"]=bTitle
            item["bAuthor"]=bAuthor
            item["bPublisher"]=bPublisher
            item["bDate"]=bDate
            item["bPrice"]=bPrice
            item["bDetail"]=bDetail
```

```
            item["bImage"]=bImage
            yield item

    def download(self,response):
        bImage = response.meta["bImage"]
        f=open("download\\"+bImage,"wb")
        f.write(response.body)
        f.close()
```

程序开始使用语句 from demo.items import BookItem 从 demo 文件夹的 items.py 文件中引入 BookItem 类的定义，在爬取到各个数据后创建 BookItem 对象 item。接下来使用语句 yield item 提交这个 item 对象，scrapy 会把这个对象推送给 pipelines.py 文件中的数据管道类去处理数据。

### 4.7.4 编写数据管道类

scrapy 框架中的 demo\demo 文件夹下的文件 pipelines.py 就是数据管道类文件，打开这个文件可以看到一个默认的数据管道类，修改并设计数据管道类如下：

```
import sqlite3
import os

class BookPipeline(object):
    def open_spider(self,spider):
        print("open_spider")
        self.con=sqlite3.connect("books.db")
        self.cursor=self.con.cursor()
        try:
            self.cursor.execute("drop table books")
        except:
            pass
        sql="create table books (ID varchar(8) primary key,bTitle varchar(256),bAuthor varchar(256),bPublisher varchar(256),bDate varchar(256),bPrice varchar(256),bDetail text,bImage varchar(256))"
        self.cursor.execute(sql)
        self.count=0
        if not os.path.exists("download"):
            os.mkdir("download")

    def close_spider(self,spider):
        print("close_spider")
        print("Total ",self.count)
        self.showDB()
        self.con.commit()
        self.con.close()

    def insertDB(self,ID,bTitle,bAuthor,bPublisher,bDate,bPrice,bDetail,bImage):
        try:
            sql="insert into books (ID,bTitle,bAuthor,bPublisher,bDate,bPrice,bDetail,bImage) values (?,?,?,?,?,?,?,?)"
            self.cursor.execute(sql,[ID,bTitle,bAuthor,bPublisher,bDate,bPrice,bDetail,bImage])
```

```
            except Exception as err:
                print(err)

    def showDB(self):
        self.cursor.execute("select ID,bTitle,bAuthor,bPublisher,
bDate,bPrice,bDetail,bImage from books")
        rows=self.cursor.fetchall()
        for row in rows:
            for r in row:
                print(r)

    def process_item(self, item, spider):
        self.count += 1
        ID = "%06d" % self.count
        bTitle=item["bTitle"]
        bAuthor=item["bAuthor"]
        bPublisher=item["bPublisher"]
        bDate=item["bDate"]
        bPrice=item["bPrice"]
        bDetail=item["bDetail"]
        bImage=item["bImage"]
        print("process_item",ID,bTitle)
        self.insertDB(ID,bTitle,bAuthor,bPublisher,bDate,bPrice,
bDetail,bImage)
        return item
```

这个类命名为 BookPipeline，它继承 object 类，类中最重要的函数是 open_spider()、process_item()与 close_spider()函数。

### 1. open_spider()函数

scrapy 开始爬取数据时会建立一个 BookPipeline 类对象，然后自动调用 open_spider()函数，因此可以在这个函数中连接数据库，建立数据库表，完成准备接收数据的基本工作。注意，在爬取数据的过程中，open_spider()函数只被调用一次，而且是在爬取开始的时候调用。

### 2. process_item()函数

当 scrapy 爬取一项数据并使用 yield item 语句提交数据时，每提交一次数据，scrapy 就会把 item 异步地推送给 BookPipeline 类的对象，并调用一次 process_item()函数，其中函数参数 item 就是推送过来的 item 对象，因此可以在这个函数中调用 insertDB()函数将数据存储到数据库中。BookPipeline 类中定义了一个类成员 self.count=0，使用它记录 process_item()函数调用的次数，并使用它产生长度为 6 位的 ID，供数据库作为关键字使用。注意，在爬取数据的过程中，process_item()函数会被反复调用，在 MySpider 的 parse()函数中有多少个 yield item 语句就调用多少次 process_item()函数，每次的 item 参数都是 yield item 语句提交的数据。

### 3. close_spider()函数

scrapy 爬取数据结束后就会自动调用 close_spider()函数，因此可以在这个函数中完成数据库的保存与关闭工作。注意，在爬取数据的过程中，close_spider()函数只被调用一次，而且是在爬取结束的时候调用。

## 4.7.5 设置 scrapy 的配置文件

scrapy 程序执行后，每爬取一个 item，就会把数据推送到 BookPipeline 类，并调用相应的 open_spider()、procees_item()、close_spider()等函数。scrapy 如何知道要执行上述操作呢？前提是必须为它设置一个通道。

在 demo\demo 文件夹中有一个 settings.py 设置文件，打开这个文件可以看到很多设置项目，大部分是用 "#" 注释的语句，找到 ITEM_PIPELINES，把它设置成如下形式：

```
# Configure item pipelines
# See http://scrapy.readthedocs.org/en/latest/topics/item-pipeline.html
ITEM_PIPELINES = {
    'demo.pipelines.BookPipeline': 300,
}
```

其中，ITEM_PIPELINES 是一个字典，把关键字改成 "demo.pipelines.BookPipeline"，而 BookPipeline 就是我们在 pipelines.py 文件中设计的数据管道类的名称，后面的 300 是一个默认的整数，它只表示顺序，实际上它可以不是 300，可以是任何整数。

这样设置后就连通了爬虫程序数据管道处理程序 pipelines.py 的通道，scrapy 工作时会把 MySpider.py 爬虫程序通过 yield item 语句提交的每项数据推送给 pipelines.py 的 BoolPipeline 类，并执行 process_item()函数，这样就可以保存数据了。

从上面的分析可以看到，scrapy 把数据爬取与数据存储分开处理，它们都是异步执行的，MySpider 每爬取到一个 item，程序就会执行 yield item 语句，将数据推送给 pipelines.py 存储，等待数据存储完毕后再爬取另外一个 item，再次执行 yield item 语句，将数据推送到 pipelines.py，然后再次存储，这个过程一直进行下去，直到爬取过程结束，数据库 books.db 中就存储了所有的爬取数据。

## 4.7.6 执行爬虫程序

从前面的分析可以看到，要使用 scrapy 的管道存储机制，必须完成下列操作：
（1）设计 items.py 中的 BookItem 类；
（2）设计 pipelines.py 中的 BookPipeline 类；
（3）设计 MySpider 类；
（4）配置 settings.py 文件。

这些操作都完成后就可以执行爬虫程序了，下面是执行的部分结果：

```
open_spider
process_item 000001 Python 编程 从入门到实践
process_item 000002 Python 编程（第四版）
process_item 000003 Python 学习手册（原书第 5 版）
process_item 000004 Python 从菜鸟到高手
process_item 000005 Python 核心编程 第 3 版
process_item 000006 Python 编程快速上手 让烦琐工作自动化（Python3 编程从入门到实践 新手学习必备用书）
process_item 000007 Python 基础教程（第 3 版）
process_item 000008 笨办法学 Python 3
process_item 000009 Python 神经网络编程
```

```
process_item 000010 Python 从小白到大牛
close_spider
Total 10
...
```

从执行的结果可以清楚地看到，程序开始时调用了 open_spider()函数，然后爬取了 10 本图书，每爬取到一本图书就调用一次 process_item()函数，爬取结束时调用了 close_spider()函数。

## 4.8 综合项目 爬取模拟图书网站数据

### 任务目标

使用 books.csv 文件中的数据，按照真实的图书网站创建一个模拟图书网站，网站包含很多网页，每个网页显示几本图书的信息。最后使用 scrapy 设计爬虫程序爬取全部网页的图书数据与图像，并将其存储到数据库。

### 4.8.1 创建模拟图书网站

#### 1. 创建网站模板

在 project4\templates 中创建一个模板文件 book.html，这个模板文件包含 bImage、bTitle、bAuthor、bPublisher、bDate、bPrice 与 bDetails 等参数，并使用 books 列表参数构造一个循环，用来显示多本图书；它还使用 pageIndex 表示当前页码，使用 pageCount 表示总页数，实现翻页功能，内容如下：

```
<style>
.pic {display:inline-block;width:200px; vertical-align:top;margin:10px;}
.info { display:inline-block; width:800px;}
.liClass { list-style: none; margin:20px;}
.title { }
.price { margin: 10px;color:red; }
.attrs { margin: 10px;color: #666; }
h3 { display:inline-block;}}
.pl {color:#888;}
.author {}
.publisher {}
.date {}
.detail {}
.link {border: 1px solid ;}
a:link { color: blue; text-decoration: none; }
a:visited { color: blue; text-decoration: none; }
</style>
<div>
<ul>
{% for b in books %}
<li class="liClass"  >
    <div class="pic">
      <img width="200"  src='/images/{{b["bImage"]}}'>
   </div>
```

```html
        <div class="info">
            <div class="title"><h3 style="display:inline-block">{{b["bTitle"]}}</h3></div>
            <div class="author">
                <span class="pl">作者</span>:<span class="attrs">{{b["bAuthor"]}}</span>
            </div>
            <div class="publisher">
                <span class="pl">出版社</span>:<span class="attrs">{{b["bPublisher"]}}</span>
            </div>
            <div class="date">
                <span class="pl">出版时间</span>:<span class="attrs">{{b["bDate"]}}</span>
            </div>
            <div class="price">
                <span class="pl">价格</span>:<span class="attrs">{{b["bPrice"]}}</span>
            </div>
            <div>简介:</div>
            <div class="detail">
                {{b["bDetail"]}}
            </div>
        </div>
    </li>
        <div></div>
{% endfor %}
</ul>
</div>
<div align="center" class="paging">
    <a href="/?pageIndex=1" class="link">第一页</a>
    {% if pageIndex>1 %}
        <a href="/?pageIndex={{pageIndex-1}}" class="link">前一页</a>
    {% else %}
        <a href="#" class="link">前一页</a>
    {% endif %}
    {% if pageIndex<pageCount %}
        <a href="/?pageIndex={{pageIndex+1}}" class="link">下一页</a>
    {% else %}
        <a href="#">下一页</a>
    {% endif %}
    <a href="/?pageIndex={{pageCount}}" class="link">末一页</a>
    <span>Page {{pageIndex}}/{{pageCount}}</span>
</div>
```

2. 创建网站服务器程序

读取 books.csv 文件中的全部图书信息并将其传递到 book.html 模板文件,形成多个网页。服务器程序 server.py 如下:

```python
import flask
app=flask.Flask(__name__,static_folder="images")

@app.route("/")
def show():
    pageRowCount=4
    if "pageIndex" in flask.request.values:
        pageIndex=int(flask.request.values.get("pageIndex"))
    else:
        pageIndex=1
    startRow=(pageIndex-1)*pageRowCount
    endRow=pageIndex*pageRowCount
    books=[]
    try:
        fobj=open("books.csv","r",encoding="utf-8")
        rows=fobj.readlines()
        count=0
        for row in rows:
            if row.strip("\n").strip()!="":
                count+=1
        count=count-1
        pageCount=count//pageRowCount
        if count % pageRowCount!=0:
            pageCount+=1
        rowIndex=0
        for i in range(1,count+1):
            row=rows[i]
            if rowIndex>=startRow and rowIndex<endRow:
                row=row.strip("\n")
                s=row.split(",")
                m={}
                m["ID"] = s[0]
                m["bImage"] = s[0] + s[7]
                m["bTitle"] = s[1]
                m["bAuthor"] = s[2]
                m["bPublisher"] = s[3]
                m["bDate"] = s[4]
                m["bPrice"] = s[5]
                m["bDetail"] = s[6]
                books.append(m)
            rowIndex+=1
        fobj.close()
    except Exception as err:
        print(err)
    return flask.render_template("book.html",books=books,
pageIndex=pageIndex,pageCount=pageCount)

app.run()
```

运行这个服务器程序，使用浏览器访问"http://127.0.0.1:5000"，就会看到图 4-8-1 所示的图书网站。

图 4-8-1 图书网站

### 4.8.2 编写数据字段类

在 items.py 文件中设计数据字段类 BookItem，如下：

```
import scrapy
class BookItem(scrapy.Item):
    # define the fields for your item here like:
    bTitle = scrapy.Field()
    bAuthor = scrapy.Field()
    bPublisher = scrapy.Field()
    bDate = scrapy.Field()
    bPrice = scrapy.Field()
    bDetail = scrapy.Field()
    bImage = scrapy.Field()
```

### 4.8.3 编写数据管道类

在 pipelines.py 文件中设计数据管道类 BookPipeline，如下：

```
import sqlite3
import os

class BookPipeline(object):
    def open_spider(self,spider):
        print("open_spider")
        self.con=sqlite3.connect("books.db")
        self.cursor=self.con.cursor()
        try:
            self.cursor.execute("drop table books")
        except:
            pass
        sql="create table books (ID varchar(8) primary key,bTitle varchar(256),bAuthor varchar(256),bPublisher varchar(256),bDate varchar(256),bPrice varchar(256),bDetail text,bImage varchar(256))"
```

```
            self.cursor.execute(sql)
            self.count=0
            if not os.path.exists("download"):
                os.mkdir("download")

        def close_spider(self,spider):
            print("close_spider")
            print("Total ",self.count)
            self.showDB()
            self.con.commit()
            self.con.close()

        def insertDB(self,ID,bTitle,bAuthor,bPublisher,bDate,bPrice,bDetail,bImage):
            try:
                sql="insert into books (ID,bTitle,bAuthor,bPublisher,bDate,bPrice,bDetail,bImage) values (?,?,?,?,?,?,?,?)"
                self.cursor.execute(sql,[ID,bTitle,bAuthor,bPublisher,bDate,bPrice,bDetail,bImage])
            except Exception as err:
                print(err)

        def showDB(self):
            self.cursor.execute("select ID,bTitle,bAuthor,bPublisher,bDate,bPrice,bDetail,bImage from books")
            rows=self.cursor.fetchall()
            for row in rows:
                for r in row:
                    print(r)

        def process_item(self, item, spider):
            self.count += 1
            ID = "%06d" % self.count
            bTitle=item["bTitle"]
            bAuthor=item["bAuthor"]
            bPublisher=item["bPublisher"]
            bDate=item["bDate"]
            bPrice=item["bPrice"]
            bDetail=item["bDetail"]
            bImage=item["bImage"]
            print(ID,bTitle)
            self.insertDB(ID,bTitle,bAuthor,bPublisher,bDate,bPrice,bDetail,bImage)
            return item
```

### 4.8.4 编写爬虫程序类

图书网站有很多个页面，爬取过程中必须实现自动翻页。很显然，翻页的按钮在<div align="center" class="paging">的<a href="…" class="link">末一页</a>中，它是一组超链接中的倒数第二个，找到这个超链接并获取href就知道下一页的地址，因此通过下面的语句获取href：

```
href=selector.xpath("//div[@class='paging']//a[position()=last()-1]/@href").extract_first().strip()
```

## 项目 ❹ 爬取图书网站数据

如果 href 不等于 "#",说明它是一个有效的地址,把它转换成绝对地址,再使用 yield 语句提交一个 scrapy.Request 对象,就实现翻页了,即:

```
if href!="#":
    href=response.urljoin(href)
    yield scrapy.Request(url=href,callback=self.parse)
```

**注意**:这实际上也是递归调用 parse()函数,只不过不是使用直接的递归调用,而是使用 scrapy 的异步机制,再次提出一个访问申请,这个申请完成访问后再次回调 parse()函数,效果与递归是一样的。

根据这个规则,修改 MySpider.py 程序,如下:

```
import scrapy
from scrapy.selector import Selector
from demo.items import BookItem

class MySpider(scrapy.Spider):
    name = "mySpider"
    def start_requests(self):
        url ='http://127.0.0.1:5000'
        yield scrapy.Request(url=url,callback=self.parse)

    def parse(self, response):
        selector=Selector(text=response.body.decode())
        lis=selector.xpath("//ul/li")
        for li in lis:
            div=li.xpath(".//div[@class='info']")
            d=div.xpath(".//div[@class='title']")
            bTitle=d.xpath("./h3/text()").extract_first()
            d=div.xpath(".//div[@class='author']")
            bAuthor=d.xpath("./span[@class='attrs']/text()").extract_first()
            d=div.xpath(".//div[@class='publisher']")
            bPublisher=d.xpath("./span[@class='attrs']/text()").extract_first()
            d=div.xpath(".//div[@class='date']")
            bDate=d.xpath("./span[@class='attrs']/text()").extract_first()
            d=div.xpath(".//div[@class='price']")
            bPrice=d.xpath("./span[@class='attrs']/text()").extract_first()
            bDetail=div.xpath(".//div[@class='detail']/text()").extract_first()
            src=li.xpath(".//div[@class='pic']//img/@src").extract_first()
            if src:
                src=response.urljoin(src).strip("/")
                p=src.rfind("/")
                bImage=src[p+1:]
                request=scrapy.Request(url=src,callback=self.download)
                request.meta["bImage"]=bImage
                yield request
            else:
```

159

```
                            bImage=""
                        #创建BookItem对象
                        item=BookItem()
                        item["bTitle"]=bTitle.strip() if bTitle else ""
                        item["bAuthor"]=bAuthor.strip() if bAuthor else ""
                        item["bPublisher"]=bPublisher.strip() if bPublisher
else ""
                        item["bDate"]=bDate.strip() if bDate else ""
                        item["bPrice"]=bPrice.strip() if bPrice else ""
                        item["bDetail"]=bDetail.strip() if bDetail else ""
                        item["bImage"]=bImage
                        yield item

            href=selector.xpath("//div[@class='paging']//a[position()=
last()-1]/@href").extract_first().strip()
            if href!="#":
                href=response.urljoin(href)
                yield scrapy.Request(url=href,callback=self.parse)

    def download(self,response):
        bImage = response.meta["bImage"]
        f=open("download\\"+bImage,"wb")
        f.write(response.body)
        f.close()
```

### 4.8.5 设置 scrapy 的配置文件

设置 settings.py 文件，如下：

```
# Configure item pipelines
# See http://scrapy.readthedocs.org/en/latest/topics/item-pipeline.html
ITEM_PIPELINES = {
    'demo.pipelines.BookPipeline': 300,
}
```

### 4.8.6 执行爬虫程序

执行爬虫程序，可以看到它爬取到了全部图书的信息，下面是部分结果：

```
open_spider
000001 Python 编程 从入门到实践
000002 Python 编程（第四版）
000003 Python 学习手册（原书第 5 版）
000004 Python 从菜鸟到高手
000005 Python 核心编程 第 3 版
000006 Python 编程快速上手 让烦琐工作自动化（Python3 编程从入门到实践 新手学习必备用书）
000007 Python 基础教程（第 3 版）
000008 笨办法学 Python 3
...
000477 深度实践 OpenStack：基于 Python 的 OpenStack 组件开发
000478 Python 程序设计教程
```

```
000479  Python 程序设计任务驱动式教程
000480  预测分析建模：Python 与 R 语言实现
close_spider
Total   480
```

## 4.9 实战项目　爬取实际图书网站数据

4-9-A 知识讲解　　4-9-B 操作演练

**任务目标**

当当网是目前国内比较大型的图书网站，本项目的目标就是爬取该网站某个主题的一类图书（如 Python 类的图书）的数据，并把爬取的数据存储到数据库中，把下载的图书图像保存到 download 文件夹中，程序综合使用了 scrapy 与 XPath 等技术。

### 4.9.1　解析网站的 HTML 代码

使用 Chrome 浏览器访问当当网，如果想知道网站上关于 Python 的图书，在搜索文本框中输入 "python" 并按 Enter 键即可，如图 4-9-1 所示。此时，相对地址转换为 "/?key=python&act=input"。Python 类的图书很多，单击下一页按钮 ">" 后相对地址转换为 "/?key=python&act=input&page_index=2"，从地址上我们知道搜索的关键字是 key 参数，当前页码是 page_index 参数，而 act=input 参数只是表明这是通过输入进行的查询。

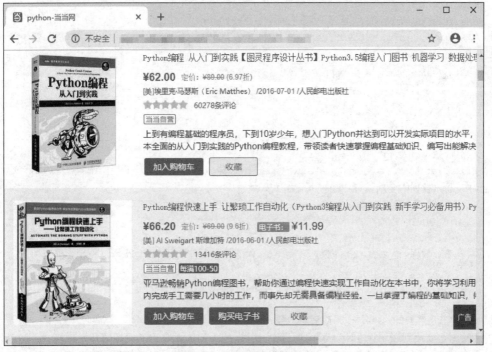

图 4-9-1　Python 类的图书

把鼠标指针移至某本图书上，单击鼠标右键，在弹出的快捷菜单中选择"检查"命令，就可以看到这本书对应的 HTML 代码，如图 4-9-2 所示。

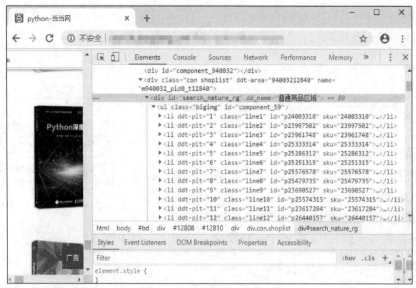

图 4-9-2  图书的 HTML 代码

仔细分析 HTML 代码的结构,我们可以看到每本图书的信息都包含在一个<li>元素中,而且它们的结构是完全一样的,这些<li>元素包含在一个<ul>元素中。复制一个<li>元素的 HTML 代码,整理后如下:

```
    <li class="line1" ddt-pit="1" id="p23473514">
      <a class="pic" dd_name="单品图片" ddclick="act=normalResult_picture&amp;pos=23473514_0_2_q" href="http://product.dangdang.com/23473514.html" name="itemlist-picture" target="_blank" title=" Python 基础教程(第 2 版·修订版) ">
        <img alt=" Python 基础教程(第 2 版·修订版) " src="http://img3x4.ddimg.cn/20/11/23473514-2_b_5.jpg"/>
      </a>
      <p class="name" name="title">
        <a dd_name="单品标题" ddclick="act=normalResult_title&amp;pos=23473514_0_2_q" href="http://product.dangdang.com/23473514.html" name="itemlist-title" target="_blank" title=" Python 基础教程(第 2 版·修订版)  Python 入门佳作 经典教程的全新修订 10 个项目引人入胜 ">
          <font class="skcolor_ljg">
          Python
          </font>
          基础教程(第 2 版·修订版)   Python 入门佳作 经典教程的全新修订 10 个项目引人入胜
        </a>
      </p>
      <p class="detail">
        本书是经典的 Python 入门教程,层次鲜明,结构严谨,内容翔实,特别是后几章,作者将前面讲述的内容应用到 10 个引人入胜的项目中,并以模板的形式介绍了项目的开发过程,手把手教授 Python 开发,让读者从项目中领略 Python 的真正魅力。本书既适合初学者夯实基础,又能帮助 Python 程序员提升技能,即使是 Python 方面的技术专家,也能从书里找到令人耳目一新的内容。
      </p>
      <p class="price">
```

```html
<span class="search_now_price">
 ¥53.00
</span>
<a class="search_discount" style="text-decoration:none;">
 定价:
</a>
<span class="search_pre_price">
 ¥79.00
</span>
<span class="search_discount">
 (6.71折)
</span>
</p>
<p class="dang" style="display: block">
 当当自营
</p>
<p class="search_star_line">
 <span class="search_star_black">
  <span style="width: 100%;">
  </span>
 </span>
 <a class="search_comment_num" dd_name="单品评论" ddclick="act=click_review_count&amp;pos=23473514_0_2_q" href="http://product.dangdang.com/23473514.html?point=comment_point" name="itemlist-review" target="_blank">
  11355条评论
 </a>
</p>
<span class="tag_box" style="background:url(http://img62.ddimg.cn/upload_img/00660/search_new_icon/jjg.png) no-repeat 0 0;_background-image:none;_filter:progid:DXImageTransform.Microsoft.AlphaImageLoader(src='http://img62.ddimg.cn/upload_img/00660/search_new_icon/jjg.png',sizingMethod='noscale');">
</span>
<p class="search_book_author">
 <span>
  (挪)
  <a dd_name="单品作者" href="/?key2=海特兰德&amp;medium=01&amp;category_path=01.00.00.00.00.00" name="itemlist-author" title="(挪)海特兰德">
   海特兰德
  </a>
 </span>
 <span>
  /2014-06-01
 </span>
 <span>
  /
  <a dd_name="单品出版社" href="/?key=&amp;key3=%C8%CB%C3%F1%D3%CA%B5%E7%B3%F6%B0%E6%C9%E7&amp;medium=01&amp;category_path=01.00.00.00.00.00" name="P_cbs" title="人民邮电出版社">
```

```
         人民邮电出版社
       </a>
      </span>
     </p>
     <div class="shop_button">
      <p class="bottom_p">
       <a class="search_btn_cart " dd_name="加入购物车" ddclick="act=
normalResult_addToCart&amp;pos=23473514_0_2_q" href="javascript:
AddToShoppingCart(23473514)" name="Buy">
         加入购物车
       </a>
       <a class="search_btn_collect" dd_name="加入收藏" ddclick="act=
normalResult_favor&amp;pos=23473514_0_2_q" href="javascript:showMsgBox
('lcase23473514','23473514','http://wish.dangdang.com/wishlist/remote_addto
favorlist.aspx');" id="lcase23473514" name="collect">
         收藏
       </a>
      </p>
     </div>
   </li>
```

### 4.9.2 爬取网站图书数据

本小节假定只关心图书的名称 bTitle、作者 bAuthor、出版时间 bDate、出版社 bPublisher、价格 bPrice、简介 bDetail 以及图书图像。

#### 1. 爬取每本图书数据

每本图书数据都包含在<li>中，如<li ddt-pit="1" class="line1"…>，每个<li>都有"ddt-pit"属性，而且 class 属性值都以"line"开始。因此使用下列语句可以获取所有的<li>，从而爬取到每本图书数据：

```
lis=selector.xpath("//li['@ddt-pit'][starts-with(@class,'line')]")
for li in lis:
      #爬取每本图书数据
```

#### 2. 爬取图书名称

从 HTML 代码可以看到，<li>中有多个<a>，图书名称包含在第一个<a>的 title 属性中，因此通过 position()=1 找出第一个<a>，然后取出 title 属性值，这个值就是图书名称。因此图书名称 bTitle 通过下列语句获取：

```
bTitle = li.xpath("./a[position()=1]/@title").extract_first()
```

#### 3. 爬取价格

价格信息包含在<li>中的 class='price'元素下面的 class='search_now_price'的<span>元素的文本中，因此价格 bPrice 通过下列语句获取：

```
bPrice = li.xpath("./p[@class='price']/span[@class='search_now_
price']/text()").extract_first()
```

#### 4. 爬取作者

作者信息包含在<li>中的 class='search_book_author'元素下面的第一个<span>元素的

title 属性中，其中，span[position()=1]用于获取第一个 <span>，因此作者 bAuthor 通过下列语句获取：

```
bAuthor = li.xpath("./p[@class='search_book_author']/span[position()=1]/a/@title").extract_first()
```

### 5. 爬取出版时间

出版时间信息包含在<li>中的 class='search_book_author'元素下面的倒数第二个<span>元素的文本中，其中，span[position()=last()-1]用于获取倒数第二个 <span>，last()用于获取最后一个<span>的序号，因此出版时间 bDate 通过下列语句获取：

```
bDate = li.xpath("./p[@class='search_book_author']/span[position()=last()-1]/text()").extract_first()
```

### 6. 爬取出版社

出版社信息包含在<li>中的 class='search_book_author'元素下面的最后一个<span>元素的 title 属性中，其中，span[position()=last()]用于获取最后一个 <span>，last()用于获取最后一个<span>的序号，因此出版社 bPublisher 通过下列语句获取：

```
bPublisher = li.xpath("./p[@class='search_book_author']/span[position()=last()]/a/@title").extract_first()
```

### 7. 爬取简介

<li>下面的 class='detail'中的文本就是图书的简介，因此简介 bDetail 通过下列语句获取：

```
bDetail = li.xpath("./p[@class='detail']/text()").extract_first()
```

### 8. 爬取图书图像

我们发现图书图像主要存储在 data-original 中，如果这里没有就存储在 src 中，如图像的 HTML 代码：

```
<a title=" Python 编程快速上手 让烦琐工作自动化（Python3 编程从入门到实践 新手学习必备用书）" ddclick="act=normalResult_picture&amp;pos=23997502_1_1_q" class="pic" name="itemlist-picture" dd_name="单品图片" href="http://product.dangdang.com/23997502.html" target="_blank">
    <img data-original="http://img3m2.ddimg.cn/1/5/23997502-1_b_6.jpg" src="http://img3m2.ddimg.cn/1/5/23997502-1_b_6.jpg" alt=" Python 编程快速上手 让烦琐工作自动化（Python3 编程从入门到实践 新手学习必备用书）" style="display: block;">
</a>
```

因此可以通过如下方式查找图像的 url：

```
url=li.xpath("./a[@name='itemlist-picture']/img/@data-original").extract_first()
    if not url:
        url = li.xpath("./a[@name='itemlist-picture']/img/@src").extract_first()
```

## 4.9.3 实现自动翻页

找到网页的翻页按钮，查看其中的 HTML 代码，如图 4-9-3 所示。最重要的是<li class="next">，其中包含下一页的地址，当跳转到最后一页时它变成<li class="next none">，如图 4-9-4 所示。

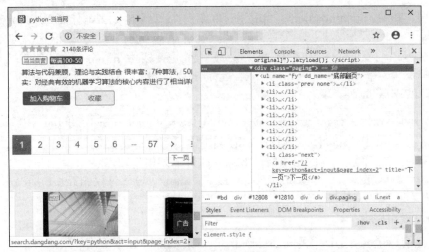

图 4-9-3　网页翻页

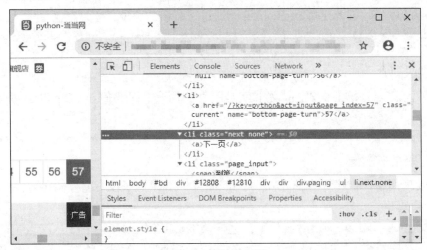

图 4-9-4　最后一页

复制<div class="paging">部分的 HTML 代码，如下：

```
    <div class="paging">
            <ul name="Fy" dd_name="底部翻页">
            <li class="prev none"><a>上一页</a></li>
            <li><a href="/?key=python&amp;act=input&amp;page_index=1" class=" current" name="bottom-page-turn">1</a></li>
            <li><a href="/?key=python&amp;act=input&amp;page_index=2" class="null" name="bottom-page-turn">2</a></li>
            <li><a href="/?key=python&amp;act=input&amp;page_index=3" class="null" name="bottom-page-turn">3</a></li>
            <li><a href="/?key=python&amp;act=input&amp;page_index=4" class="null" name="bottom-page-turn">4</a></li>
            <li><a href="/?key=python&amp;act=input&amp;page_index=5" class="null" name="bottom-page-turn">5</a></li>
            <li><a href="/?key=python&amp;act=input&amp;page_index=6" class="null" name="bottom-page-turn">6</a></li>
            <li><span>...</span></li>
```

```
            <li><a href="/?key=python&amp;act=input&amp;page_index=57"
class="null" name="bottom-page-turn">57</a></li>
            <li class="next"><a href="/?key=python&amp;act=input&amp;
page_index=2" title="下一页">下一页</a></li>
            <li class="page_input">
                <span>到第</span> <input id="t__cp" type="text"
class="number" value="1">
                <span>页</span>
                <input class="button" id="click_get_page" value=
"确定" type="button" name="bottom-page-turn">
            </li>
        </ul>
    </div>
```

因此可以通过如下方式找到翻页地址 url 并实现翻页：

```
link=selector.xpath("//div[@class='paging']/ul[@name='Fy']/li[@class=
'next']/a/@href").extract_first()
if link:
    url=response.urljoin(link)
    yield scrapy.Request(url=url, callback=self.parse)
```

如果是最后一页，则没有<li class="next">元素，即当 link 为 None 时，翻页停止。

### 4.9.4 编写爬虫程序

**1. 编写 items.py 中的数据字段类**

items.py 中的数据字段类 BookItem 编写如下：

```
import scrapy
class BookItem(scrapy.Item):
    # define the fields for your item here like:
    ID=scrapy.Field()
    bTitle = scrapy.Field()
    bAuthor = scrapy.Field()
    bPublisher = scrapy.Field()
    bDate = scrapy.Field()
    bPrice = scrapy.Field()
    bDetail = scrapy.Field()
    bImage = scrapy.Field()
```

**2. 编写 pipelines.py 中的数据管道类**

pipelines.py 中的数据管道类 BookPipeline 编写如下：

```
import sqlite3
import os

class BookPipeline(object):
    def open_spider(self,spider):
        print("open_spider")
        self.con=sqlite3.connect("books.db")
        self.cursor=self.con.cursor()
        try:
            self.cursor.execute("drop table books")
        except:
```

```
                pass
            sql="create table books (ID varchar(8) primary key,bTitle
varchar(256),bAuthor varchar(256),bPublisher varchar(256),bDate varchar(256),
bPrice varchar(256),bDetail text,bImage varchar(256))"
            self.cursor.execute(sql)
            if not os.path.exists("download"):
                os.mkdir("download")

    def close_spider(self,spider):
        print("close_spider")
        print("Total ",spider.count)
        self.con.commit()
        self.con.close()

    def insertDB(self,ID,bTitle,bAuthor,bPublisher,bDate,bPrice,
bDetail,bImage):
        try:
            sql="insert into books (ID,bTitle,bAuthor,bPublisher,
bDate,bPrice,bDetail,bImage) values (?,?,?,?,?,?,?,?)"
            self.cursor.execute(sql,[ID,bTitle,bAuthor,bPublisher,
bDate,bPrice,bDetail,bImage])
        except Exception as err:
            print(err)

    def process_item(self, item, spider):
        ID=item["ID"]
        bTitle=item["bTitle"]
        bAuthor=item["bAuthor"]
        bPublisher=item["bPublisher"]
        bDate=item["bDate"]
        bPrice=item["bPrice"]
        bDetail=item["bDetail"]
        bImage=item["bImage"]
        print(ID,bTitle)
        self.insertDB(ID,bTitle,bAuthor,bPublisher,bDate,bPrice,
bDetail,bImage)
        return item
```

在爬取的过程中，一旦打开一个 spider 爬虫程序，就会执行这个类的 open_spider(self, spider)函数；一旦这个 spider 爬虫程序关闭，就执行这个类的 close_spider(self,spider)函数。因此程序在 open_spider()函数中连接数据库并创建操作游标 self.cursor，在 close_spider()函数中提交数据库并关闭数据库。程序中使用 spider.count 变量统计爬取的图书数量。

在数据处理函数中，每当有数据到达时就显示图书的 ID 与名称，并使用 insert 的 SQL 语句把数据插入数据库。

3. 编写 scrapy 的配置文件

scrapy 的配置文件 settings.py 中要添加语句：

```
ITEM_PIPELINES = {
    'demo.pipelines.BookPipeline': 300,
}
```

这样就可以把爬取的数据推送到 pipelines 的 BookPipeline 类中。

### 4. 编写爬虫程序

根据前面的 HTML 代码分析，我们可以编写爬虫程序 MySpider.py 如下：

```
import scrapy
from demo.items import BookItem

class MySpider(scrapy.Spider):
    name = "mySpider"
    key = 'python'
    source_url='http://search.dangdang.com/'

    def start_requests(self):
        self.count=0
        url = MySpider.source_url+"?key="+MySpider.key
        yield scrapy.Request(url=url,callback=self.parse)

    def getFileExt(self, url):
        p = url.rindex(".")
        if p >= 0:
            s = url[p:]
            if s[len(s) - 1] == "/":
                s = s[0:len(s) - 1]
        else:
            s = ""
        return s

    def parse(self, response):
        try:
            selector=scrapy.Selector(text=response.body.decode("gbk"))
            lis=selector.xpath("//li['@ddt-pit'][starts-with(@class,'line')]")
            for li in lis:
                self.count+=1
                ID="%06d" %self.count
                bTitle=li.xpath("./a[position()=1]/@title").extract_first()
                bPrice = li.xpath("./p[@class='price']/span[@class='search_now_price']/text()").extract_first()
                bAuthor = li.xpath("./p[@class='search_book_author']/span[position()=1]/a/@title").extract_first()
                bDate = li.xpath("./p[@class='search_book_author']/span[position()=last()-1]/text()").extract_first()
                bPublisher = li.xpath("./p[@class='search_book_author']/span[position()=last()]/a/@title").extract_first()
                bDetail = li.xpath("./p[@class='detail']/text()").extract_first()
                url=li.xpath("./a[@name='itemlist-picture']/img/@data-original").extract_first()
                if not url:
                    url = li.xpath("./a[@name='itemlist-picture']/img/@src").extract_first()
                if url:
```

```
                              #下载图像
                              bImage=ID+self.getFileExt(url)
                              url=response.urljoin(url)
                              request=scrapy.Request(url=url,callback=self.download)
                              request.meta["bImage"]=bImage
                              yield request
                         else:
                              bImage=""
          #如果某个数据字段不存在，则extract_first()返回None，并把数据设置为空字符串
                         item=BookItem()
                         item["ID"]=ID
                         item["bTitle"]=bTitle.strip() if bTitle else ""
                         item["bAuthor"]=bAuthor.strip() if bAuthor else ""
                         item["bDate"] = bDate.strip()[1:] if bDate else ""
                         item["bPublisher"] = bPublisher.strip() if bPublisher else ""
                         item["bPrice"] = bPrice.strip() if bPrice else ""
                         item["bDetail"] = bDetail.strip() if bDetail else ""
                         item["bImage"]=bImage
                         yield item
                #最后一页时，link为None
                    link=selector.xpath("//div[@class='paging']/ul[@name='Fy']/li[@class='next']/a/@href").extract_first()
                    if link:
                         url=response.urljoin(link)
                         yield scrapy.Request(url=url, callback=self.parse)
          except Exception as err:
                    print(err)

     def download(self,response):
          bImage = response.meta["bImage"]
          try:
                f=open("download\\"+bImage,"wb")
                f.write(response.body)
                f.close()
                print("downloaded",bImage)
          except:
                print("downloading",bImage,"failed")
```

这个程序在start_requests()函数中定义了一个self.count变量，它是这个爬虫类MySpider实例的变量，使用它统计爬取的图书数量，并生成一个长度为6位的ID供数据库使用。同时这个变量在BookPipeline类的close_spider()函数中再次被用到，因为该函数的spider参数就是这个MySpider类的实例，所以close_spider()函数中的spider.count就是MySpider中的self.count。

### 4.9.5 执行爬虫程序并查看爬取结果

#### 1. 执行爬虫程序

执行爬虫程序就可以爬取到所有Python类的图书数据，并将这些图书的数据存储到books.db数据库中，下面是部分结果：

```
open_spider
000001 Python 编程 从入门到实践
000002 Python 编程快速上手 让烦琐工作自动化（Python3 编程从入门到实践 新手学习必备用书）
000003 Python 核心编程 第 3 版
000004 Python 从菜鸟到高手
000005 笨办法学 Python 3
000006 Python 神经网络编程
000007 Python 学习手册（原书第 5 版）
000008 Python 编程从零基础到项目实战（微课视频版）
...
downloaded 003362.jpg
downloaded 003365.jpg
downloaded 003363.jpg
downloaded 003367.jpg
downloaded 003366.jpg
close_spider
Total  3408
```

## 2. 查看爬取结果

（1）查看数据库

编写一个简单的查看程序 show.py 来查看 books.db 中的内容。执行下面的程序可以看到数据库中存储了 3408 本图书：

```python
import sqlite3
con=sqlite3.connect("books.db")
cursor=con.cursor()
cursor.execute("select ID,bTitle,bAuthor,bPublisher,bDate,bPrice,bDetail,bImage from books")
rows = cursor.fetchall()
for row in rows:
    for r in row:
        print(r)
print("Total ",len(rows))
con.close()
```

（2）查看图书图像

在 download 文件夹中可以看到爬取了 3408 幅图像，如图 4-9-5 所示。

图 4-9-5　爬取的图像

# Python 爬虫项目教程（微课版）

## 项目总结

本项目涉及一个包含多个网页的图书网站，我们使用 scrapy 爬取各个网页的数据，实现了爬取图书网站数据的爬虫程序。

scrapy 是一个优秀的分布式爬取框架，它规定了一系列的程序规则，例如，使用 items.py 定义数据的格式，使用 pipelines.py 实现数据的存储，使用 spider.py 实现数据的爬取，使用 settings.py 规定各个模块之间的联系，对复杂的爬虫程序进行模块化管理。我们只需按规则填写各个模块即可，各个模块的协调工作由 scrapy 自动完成。而且 scrapy 支持使用 XPath 与 CSS 方法查找网页数据。使用 scrapy 可以高效地爬取大多数网站的数据。

但是实际上有些网站的数据是使用 JavaScript 管理的，一般的 scrapy 并不能执行 JavaScript 程序，在后面的项目中将介绍能执行 JavaScript 程序的爬虫程序技术。

## 练习 4

1. 简单说明 scrapy 的入口地址的规定。
2. 比较使用 XPath 和 BeautifulSoup 解析 HTML 代码的特点与区别。
3. 在 scrapy 中如何把爬取的数据写入数据库？items.py 与 pipelines.py 文件有什么作用？
4. Python 中的 yield 语句如何工作？为什么 scrapy 爬到的数据使用 yield 语句返回，而不是使用 return 语句返回？
5. scrapy 是如何实现爬取多个网页的数据的？如何理解分布式爬取过程？
6. 有一个服务器程序，它根据请求页面的不同来展示 3 组学生信息，这些学生信息分别存储在 students1.txt、students2.txt、students3.txt 文件中，每个页面有相同的结构（学号 No、姓名 Name、性别 Gender、年龄 Age），参数 page=N 控制展示 studentsN.txt 的学生表格（N=1,2,3），程序如下：

```
import flask
app=flask.Flask(__name__)
@app.route("/")
def show():
    page = flask.request.args.get("page") if "page" in flask.request.args else "1"
    maxpage=3
    page=int(page)
    st="<h3>学生信息表</h3>"
    st=st+"<table border='1' width='300'>"
    fobj=open("students"+str(page)+".txt","rt",encoding="utf-8")
    while True:
        #读取一行，去除行尾部"\n"换行符号
        s=fobj.readline().strip("\n")
        #如果读到文件尾部就退出
        if s=="":
            break
        #按逗号拆分
        s=s.split(",")
```

```
                st=st+"<tr>"
                #把各个数据组织在<td>…</td>中
                for i in range(len(s)):
                        st=st+"<td>"+s[i]+"</td>"
                #完成一行
                st=st+"</tr>"
        fobj.close()
        st=st+"</table>"
        st=st+"<div>"
        if page>1:
            st = st + "<a href='/?page=" +str(page-1) + "'>【前一页】</a>"
        if page<maxpage:
            st=st+"<a href='/?page="+str(page+1)+"'>【下一页】</a>"
        st=st+"</div>"
        st=st+"<input type='hidden' name='page' value='"+str(page)+"'>"
        st=st+"<div>Page: "+str(page)+"/"+str(maxpage)+"</div>"
        return st

if __name__=="__main__":
    app.run()
```

执行服务器程序,学生信息表如图 4-11-1 所示。

图 4-11-1　学生信息表

使用 scrapy 设计一个爬虫程序,爬取所有学生的信息并将其存储到数据库中。

7. 中国是诗的国度,唐诗、宋词、元曲是我们祖先智慧的结晶,是中国文化最灿烂的瑰宝之一,编写爬虫程序爬取图书网站中的中国古诗词书籍。

# 项目 5 爬取商城网站数据

拓展阅读

"中国芯"的飞天之路

实际中很多网站的网页都不是静态的 HTML 文档，很多信息都是通过 JavaScript 程序处理后才显示出来的动态数据，使用普通的爬虫程序不能爬取这些动态数据。Selenium 就是这样一种能模拟浏览器执行 JavaScript 程序的工具。本项目介绍如何使用 Selenium 编写爬虫程序来爬取商城网站动态手机数据。

手机是我们日常生活必备的物品，手机的核心部件是 CPU（中央处理器），高端的 CPU 技术非常复杂，一直被国外垄断。我的中国芯，我的中国梦，我国要有自主知识产权的CPU，才能杜绝被国外卡脖子。

5-1-A

知识讲解

5-1-B

操作演练

## 5.1 项目任务

使用浏览器访问京东商城网站，在搜索文本框中输入"手机"后按 Enter 键，可以看到图 5-1-1 所示的页面。分析京东商城的网页发现，很多数据是由 JavaScript 程序控制的。在这个项目中将使用 selenium 设计一个爬虫程序来爬取所有的手机数据与图像。

图 5-1-1 京东商城网站

在爬取京东商城网站数据之前先学习爬取模拟商城网站的数据。创建项目文件夹 project5，在 project5 文件夹中有一个 phones.csv 文件，文件中存储了手机的数据，其中前面几行数据如下：

```
ID,mMark,mPrice,mNote,mImage
000001,荣耀 9i,1198.0,荣耀 9i 4GB+64GB 幻夜黑 移动联通电信 4G 全面屏手机 双卡双
```

待,000001.jpg
　　000002,荣耀 8X,1399.0,荣耀 8X 千元屏霸 91%屏占比 2000 万 AI 双摄 4GB+64GB 幻夜黑 移动联通电信 4G 全面屏 双卡双待,000002.jpg
　　000003,小米 8,2299.0,小米 8 全面屏游戏智能手机 6GB+64GB 蓝色 全网通 4G 双卡双待,000003.jpg
　　……

除第 1 行外，其中每行都包含某部手机的数据，各个字段之间用","分开，第 1 个字段是编号，第 2 个字段是手机品牌，第 3 个字段是手机价格，第 4 个字段是手机简介，第 5 个字段是图像名称。project5\images 文件夹中存储了每部手机的图像，如图 5-1-2 所示。

图 5-1-2　手机图像

使用 phones.csv 文件中的数据建立一个模拟商城网站，如图 5-1-3 所示。网页的很多数据是由 JavaScript 程序控制的，例如，各个翻页按钮由<input type='button'>创建，单击按钮时执行对应的 JavaScript 函数实现翻页。本项目将使用 selenium 设计一个爬虫程序来爬取这个模拟商城网站中所有的手机数据与图像。

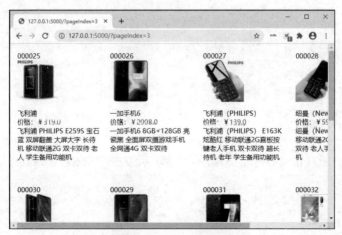

图 5-1-3　模拟商城网站

## 5.2 使用 selenium 编写爬虫程序

5-2-A 知识讲解

5-2-B 操作演练

**任务目标**

现在很多网站的网页中的数据都是通过 JavaScript 程序来控制的,那么如何让程序模拟浏览器执行 JavaScript 程序呢?selenium 能实现这个功能。在本节中,我们主要学习 selenium 环境的搭建,并使用 selenium 编写爬虫程序。

### 5.2.1 JavaScript 程序控制网页

网页上的信息不一定都是静态的 HTML 数据,实际上很多信息都是通过 JavaScript 程序处理后得到的,那么如何爬取这些数据呢?我们先来设计一个包含 JavaScript 程序的网页文档,看看如何爬取它的数据。

#### 1. 创建网站模板

在 project5\templates 文件夹中设计一个模板文件 phone.html,它包含 3 个<span>:第 1 个<span>的 id 是 "hMsg",它的信息是确定的 "Html Message";第 2 个<span>的 id 是 "jMsg",它的信息是在网页加载时由 JavaScript 的程序赋予的值 "javascript Message";第 3 个<span>的 id 是 "sMsg",它的信息是在网页加载时通过 Ajax 的方法向服务器提出 GET 请求获取的,服务器返回字符串值 "Server Message"。phone.html 模板文件如下:

```
<script>
    function init()
    {
        http=new XMLHttpRequest();
        http.open("get","/show",false);
        http.send(null);
        msg=http.responseText;
        document.getElementById("sMsg").innerHTML=msg;
        document.getElementById("jMsg").innerHTML="javascript Message";
    }
</script>
<body onload="init()">
Testing<br>
<span id="hMsg">Html Message</span><br>
<span id="jMsg"></span><br>
<span id="sMsg"></span>
</body>
```

#### 2. 创建网站服务器程序

服务器程序 server.py 显示出 phone.html 文件的内容,其中,index()函数读取该文件并发送出去,show()函数在接收地址 "/show" 请求后发送 "Server Message"。

```
import flask
app=flask.Flask(__name__)
@app.route("/")
def index():
    return flask.render_template("phone.html")
```

```
@app.route("/show")
def show():
    return "Server Message"
app.run()
```

### 3. 使用浏览器访问

运行服务器程序 server.py，使用浏览器访问"http://127.0.0.1:5000"，结果如图 5-2-1 所示。

图 5-2-1　测试网站

## 5.2.2 普通爬虫程序的问题

### 1. 编写普通客户端程序

编写一个爬虫程序 spider.py，它通过 urllib.request 直接访问"http://127.0.0.1:5000"，程序如下：

```
import urllib.request
resp=urllib.request.urlopen("http://127.0.0.1:5000");
data=resp.read()
data=data.decode()
print(data)
```

执行该程序，我们看到输出的结果如下：

```
<script>
    function init()
    {
        http=new XMLHttpRequest();
        http.open("get","/show",false);
        http.send(null);
        msg=http.responseText;
        document.getElementById("sMsg").innerHTML=msg;
        document.getElementById("jMsg").innerHTML="javascript Message";
    }
</script>
<body onload="init()">
<span id="hMsg">Html Message</span><br>
<span id="jMsg"></span><br>
<span id="sMsg"></span>
</body>
```

这个结果就是 phone.html 文件。注意，在输出的结果中没有 id="jMsg"与 id="sMsg"的 <span>的信息，这些信息在程序运行后产生。

### 2. 编写普通爬虫程序

编写一个爬虫程序 spider.py，它通过 urllib.request 直接访问"http://127.0.0.1:5000"来获取 HTML 代码，使用 BeautifulSoup 解析得到数据，程序如下：

```python
from bs4 import BeautifulSoup
import urllib.request
resp=urllib.request.urlopen("http://127.0.0.1:5000");
html=resp.read()
html=html.decode()
soup=BeautifulSoup(html,"lxml")
hMsg=soup.find("span",attrs={"id":"hMsg"}).text
print(hMsg)
jMsg=soup.find("span",attrs={"id":"jMsg"}).text
print(jMsg)
sMsg=soup.find("span",attrs={"id":"sMsg"}).text
print(sMsg)
```

执行该程序，结果如下：

```
Html Message
```

显然，如果通过该方法获取网页的 HTML 代码并进行数据爬取，那么只能爬取 hMsg 的信息"Html Message"，但是爬取不到 jMsg 与 sMsg 的信息。

因为 jMsg 与 sMsg 的信息不是静态地嵌入在网页中的，而是通过 JavaScript 与 Ajax 动态产生的。通过 urllib.request.urlopen 得到的网页中没有这样的动态信息，如果要得到这些信息，就必须让爬虫程序能够执行对应的 JsvaScript 程序，selenium 框架能实现这个功能。

### 5.2.3 安装 selenium 与 Chrome 驱动程序

前面已经分析了要获取 jMsg 与 sMsg 的信息就必须使客户端在获取网页后能按要求执行对应的 JavaScript 程序。显然，一般的客户端程序没有这个能力去执行 JavaScript 程序，因此必须寻找一个能像浏览器那样工作的插件来完成这个工作，它就是 selenium。selenium 是一个没有显示界面的浏览器，它能与通用的浏览器（如 Chrome、Firefox 等）配合工作。下面安装 selenium 与 Chrome 的驱动程序。

（1）安装 selenium

执行以下命令即可安装 selenium。

```
pip install selenium
```

（2）安装 Chrome 的驱动程序

要想使 selenium 与浏览器配合工作，就必须安装浏览器对应的驱动程序。例如，要与 Chrome 配合工作，就要先下载 chromedrive.exe 的驱动程序，然后把它复制到 Python 的 scripts 目录下。

### 5.2.4 编写 selenium 爬虫程序

#### 1. 使用 selenium 获取网页的 HTML 代码

按下列步骤编写客户端程序。

（1）程序先从 selenium 引入 webdriver，并引入 Chrome 的驱动程序的选择项目 Options：

```python
from selenium import webdriver
from selenium.webdriver.chrome.options import Options
```

## 项目 ❺ 爬取商城网站数据

（2）设置启动 Chrome 浏览器时不可见：
```
chrome_options.add_argument('--headless')
chrome_options.add_argument('--disable-gpu')
```
（3）创建 Chrome 浏览器：
```
driver= webdriver.Chrome(chrome_options=chrome_options)
```
这样创建的 Chrome 浏览器是不可见的，如果仅仅使用：
```
driver= webdriver.Chrome()
```
创建 Chrome 浏览器，那么在程序执行时会弹出一个 Chrome 浏览器。

（4）使用 driver.get(url)函数访问网页：
```
driver.get("http://127.0.0.1:5000")
```
（5）通过 driver.page_source 获取网页的 HTML 代码：
```
html=driver.page_source
print(html)
```
（6）使用 driver.close()函数关闭浏览器：
```
driver.close()
```
根据这样的规则编写爬虫程序 spider.py，如下：
```
from selenium import webdriver
from selenium.webdriver.chrome.options import Options
chrome_options = Options()
chrome_options.add_argument('--headless')
chrome_options.add_argument('--disable-gpu')
driver= webdriver.Chrome(chrome_options=chrome_options)
driver.get("http://127.0.0.1:5000")
html=driver.page_source
print(html)
driver.close()
```
执行该程序，结果如下：
```
<html xmlns="http://www.w3.org/1999/xhtml">
<head><script>
    function init()
    {
        http=new XMLHttpRequest();
        http.open("get","/show",false);
        http.send(null);
        msg=http.responseText;
        document.getElementById("sMsg").innerHTML=msg;
        document.getElementById("jMsg").innerHTML="javascript Message";
    }
</script></head>
<body onload="init()">
<span id="hMsg">Html Message</span><br />
<span id="jMsg">javascript Message</span><br />
<span id="sMsg">Server Message</span>
</body>
</html>
```
由此可见，我们得到的 HTML 代码是执行完 JavaScript 程序后的代码，其中包含 jMsg 与 sMsg 的信息。

### 2. 编写 selenium 爬虫程序

selenium 模拟浏览器访问网站的方法来获取网页的 HTML 代码，然后从中爬取需要的数据，这样的爬虫程序功能就比较强大了。编写爬虫程序 spider.py，如下：

```python
from selenium import webdriver
from selenium.webdriver.chrome.options import Options
from bs4 import BeautifulSoup
chrome_options = Options()
chrome_options.add_argument('--headless')
chrome_options.add_argument('--disable-gpu')
driver = webdriver.Chrome(chrome_options=chrome_options)
driver.get("http://127.0.0.1:5000")
html=driver.page_source
soup=BeautifulSoup(html,"lxml")
hMsg=soup.find("span",attrs={"id":"hMsg"}).text
print(hMsg)
jMsg=soup.find("span",attrs={"id":"jMsg"}).text
print(jMsg)
sMsg=soup.find("span",attrs={"id":"sMsg"}).text
print(sMsg)
```

执行该程序后，我们爬取到了所有数据：

```
Html Message
javascript Message
Server Message
```

由此可见，selenium 主要是模拟浏览器去访问网页，并充分执行网页中的 JavaScript 程序，使得网页中的数据被充分下载，这样再用爬虫程序去爬取数据就比较稳妥了。

## 5.3 使用 selenium 查找 HTML 元素

5-3-A 知识讲解

5-3-B 操作演练

**任务目标**

获取了网页的 HTML 代码后就可以使用很多方法查找 HTML 元素并爬取其中的数据。selenium 支持 XPath、CSS 等多种查找元素的方法，掌握这些方法可以灵活地爬取所需的数据。在本节中，我们主要学习使用 selenium 的 XPath、CSS 等方法查找元素。

### 5.3.1 创建模拟商城网站

#### 1. 创建网站模板

在 templates 文件夹中创建模板文件 phone.html，内容如下：

```
<style>
.pic {display:inline-block;width:200px; vertical-align:top;margin:10px;}
.info { display:inline-block; width:500px;}
.title { }
.price { margin: 10px;color:red; }
h3 { display:inline-block;}}
.pl {color:#888;}
```

```html
        .author {}
        .date {}
        .detail {}
    </style>
    <div>
        <div class="pic">
            <img width="200" id="image" src='/images/000001.jpg'>
        </div>
        <div class="info">
            <div class="title"><h3 id="title" style="display:inline-block">荣耀9i</h3></div>
            <div class="mark">
                <span class="pl">品牌</span>:<span name="mark">华为</span>
            </div>
            <div class="date">
                <span class="pl">生产日期</span>:<span name="date">2016-12-01</span>
            </div>
            <div class="price">
                <span class="pl">价格</span>:<span name="price">¥1200.00</span>
            </div>
            <div>简介:</div>
            <div class="detail">
                荣耀9i 4GB+64GB 幻夜黑 <a href="#">移动联通</a>电信4G全面屏手机<a href="#">双卡双待</a>
            </div>
        </div>
    </div>
```

## 2. 创建网站服务器程序

创建服务器程序 server.py 来展示 phone.html 网页，程序如下：

```python
import flask
app=flask.Flask(__name__,static_folder="images")
@app.route("/")
def show():
    return flask.render_template("phone.html")
app.run()
```

执行这个服务器程序，使用浏览器访问 "http:/127.0.0.1:5000"，结果如图 5-3-1 所示。

图 5-3-1　商城网站

### 3. 创建查找程序

selenium 支持多种查找 HTML 元素的方法，通过这些方法就可以爬取到所需的数据。为了方便说明，先设计下列程序：

```
from selenium import webdriver
from selenium.webdriver.chrome.options import Options
chrome_options = Options()
chrome_options.add_argument('--headless')
chrome_options.add_argument('--disable-gpu')
driver= webdriver.Chrome(chrome_options=chrome_options)
driver.get("http://127.0.0.1:5000")
#查找元素与数据
driver.close()
```

其中，"#查找元素与数据"部分就是使用各种方法查找元素与数据的地方。

## 5.3.2 使用 XPath 查找元素

使用 XPath 查找元素主要涉及以下两个函数。

（1）find_element_by_xpath(xpath)函数：查找 xpath 匹配的第一个元素，如果找到就返回一个 WebElement 对象，如果找不到就抛出异常。

（2）find_elements_by_xpath(xpath)函数：查找 xpath 匹配的所有元素组成的列表，每个元素都是一个 WebElement 对象，如果找不到就返回空列表。

任何一个 WebElement 对象都可以调用 find_element_by_xpath()与 find_elements_by_xpath()函数。

**例 5-3-1**：查找网页中的<h3>元素。

```
elem=driver.find_element_by_xpath("//div[@class='info']//h3")
print(type(elem))
```

结果：

```
<class 'selenium.webdriver.remote.webelement.WebElement'>
```

**例 5-3-2**：查找网页中的<h4>元素。

```
try:
    elem=driver.find_element_by_xpath("//div[@class='info']//h4")
    print(type(elem))
except Exception as err:
    print(err)
```

没有找到<h4>，抛出下列异常：

```
Message: no such element: Unable to locate element: {"method":"xpath","selector":"//div[@class='info']//h4"}
```

## 5.3.3 查找元素的文本与属性

通过 WebElement 对象可以查找元素的文本与属性。

（1）任何一个 WebElement 对象都可以通过 text 属性获取它的文本，元素的文本值是它与它的所有子孙节点的文本的组合，如果没有就返回空字符串。

（2）任何一个 WebElement 对象都可以通过 get_attribute(attrName)获取名称为 attrName 的属性值，如果元素没有 attrName 属性就返回 None。

**例 5-3-3**：查找网页中第一个<span>元素的文本。

```
print(driver.find_element_by_xpath("//div[@class='info']//span").text)
```

结果：

```
品牌
```

**例 5-3-4**：查找网页中<div class='mark'>中的所有<span>元素的文本。

```
elem=driver.find_element_by_xpath("//div[@class='mark']")
elems=elem.find_elements_by_xpath(".//span")
for elem in elems:
    print(elem.text)
```

结果：

```
品牌
华为
```

**例 5-3-5**：查找网页中手机的品牌。

```
print(driver.find_element_by_xpath("//div[@class='info']//span[@name='mark']").text)
```

结果：

```
华为
```

**例 5-3-6**：查找网页中手机图像<img>的地址。

```
print(driver.find_element_by_xpath("//div[@class='pic']//img").get_attribute("src"))
```

结果：

```
http://127.0.0.1:5000/images/000001.jpg
```

**例 5-3-7**：查找网页中手机图像<img>的 alt 属性与 xxx 属性。

```
elem=driver.find_element_by_xpath("//div[@class='pic']//img").get_attribute("alt")
print(type(elem),len(elem))
elem=driver.find_element_by_xpath("//div[@class='pic']//img").get_attribute("xxx")
print(elem)
```

结果：

```
<class 'str'> 0
None
```

值得注意的是，<img>默认有 alt 属性，只是这个网页中没有设置，因此获取的 alt 属性值是空字符串；但是<img>默认没有 xxx 属性，因此得到 None。

**例 5-3-8**：查找网页中<div class='mark'>的 HTML 文本。

```
elem=driver.find_element_by_xpath("//div[@class='mark']")
print("innerHTML")
print(elem.get_attribute("innerHTML").strip())
print("outerHTML")
print(elem.get_attribute("outerHTML").strip())
```

结果：

```
innerHTML
<span class="pl">品牌</span>:<span name="mark">华为</span>
outerHTML
<div class="mark">
```

```
    <span class="pl">品牌</span>:<span name="mark">华为</span>
</div>
```

### 5.3.4 使用 id 值查找元素

HTML 中的很多元素都有唯一的 id 值，selenium 可以通过 id 值查找元素。

driver.find_element_by_id(id)函数：查找 id 匹配的第一个元素，如果找到就返回一个 WebElement 对象，如果没有找到就抛出异常。

**例 5-3-9**：查找网页中 id="title"的元素的文本。

```
print(driver.find_element_by_id("title").text)
```

结果：

```
荣耀9i
```

**例 5-3-10**：查找网页中 id="name"的元素。

```
try:
    print(driver.find_element_by_id("name"))
except Exception as err:
    print(err)
```

结果抛出一个异常：

```
Message: no such element: Unable to locate element: {"method":"id","selector":"name"}
```

### 5.3.5 使用 name 属性值查找元素

HTML 中的很多元素都有一个 name 属性值，selenium 可以通过 name 属性值查找元素。

（1）find_element_by_name(value)函数：查找 name=value 匹配的第一个元素，如果找到就返回一个 WebElement 对象，如果找不到就抛出异常。

（2）find_elements_by_name(value)函数：查找 name=value 匹配的所有元素组成的列表，每个元素都是一个 WebElement 对象，如果找不到就返回空列表。

**例 5-3-11**：查找网页中手机的品牌。

```
print(driver.find_element_by_name("mark").text)
```

结果：

```
华为
```

**例 5-3-12**：查找网页中 name="xxx"的元素。

```
try:
    driver.find_element_by_name("xxx")
except Exception as err:
    print(err)
```

结果抛出一个异常：

```
Message: no such element: Unable to locate element: {"method":"name","selector":"xxx"}
```

### 5.3.6 使用 CSS 查找元素

selenium 也支持使用 CSS 查找元素。

（1）find_element_by_css_selector(css)函数：查找 css 匹配的第一个元素，如果找到就

返回一个 WebElement 对象，如果找不到就抛出异常。

（2）find_elements_by_css_selector(css)函数：查找 css 匹配的所有元素组成的列表，每个元素都是一个 WebElement 对象，如果找不到就返回空列表。

**例 5-3-13**：查找网页中手机的品牌。
```
print(driver.find_element_by_css_selector("div[class='info'] span[name='mark']").text)
```
结果：
```
华为
```

**例 5-3-14**：查找网页中手机图像的地址。
```
print(driver.find_element_by_css_selector("div[class='mark']>div").get_attribute("src"))
```
结果：
```
http://127.0.0.1:5000/images/000001.jpg
```

**例 5-3-15**：查找网页中<div class='mark'>下面的所有元素。
```
elems=driver.find_elements_by_css_selector("div[class='mark'] *")
for elem in elems:
    print(elem.text)
```
结果：
```
品牌
华为
```

**例 5-3-16**：查找网页中手机的型号。
```
print(driver.find_element_by_css_selector("#title").text)
```
或者
```
print(driver.find_element_by_css_selector("[id='title']").text)
```
结果：
```
荣耀 9i
```

### 5.3.7 使用 tagName 查找元素

selenium 还可以使用 HTML tagName 查找元素。

find_elements_by_tag_name(tagName)函数：查找 tagName 匹配的所有元素，如果找到就返回一个 WebElement 列表，如果找不到就返回空列表。

**例 5-3-17**：查找网页中<div class='mark'>元素下面的所有<span>元素。
```
elem=driver.find_element_by_xpath("//div[@class='mark']")
elems=elem.find_elements_by_tag_name("span")
for elem in elems:
    print(elem.text)
```
结果：
```
品牌
荣耀
```

**例 5-3-18**：查找网页中手机的型号。
```
print(driver.find_element_by_tag_name("h3").text)
```
结果：
```
荣耀 9i
```

## 5.3.8 使用文本查找超链接

selenium 可以通过超链接的文本来查找该超链接。

（1）find_element_by_link_text(text)函数：查找第一个文本值为 text 的超链接元素<a>，如果找到就返回该元素的 WebElement 对象，如果找不到就抛出异常。

（2）find_element_by_partial_link_text(text)函数：查找第一个文本值包含 text 的超链接元素<a>，如果找到就返回该元素的 WebElement 对象，如果找不到就抛出异常。

（3）find_elements_by_link_text(text)函数：查找所有文本值为 text 的超链接元素<a>，如果找到就返回 WebElement 列表，如果找不到就返回空列表。

（4）find_elements_by_partial_link_text(text)函数：查找所有文本值包含 text 的超链接元素<a>，如果找到就返回 WebElement 列表，如果找不到就返回空列表。

**例 5-3-19**：查找网页中<a href="#">移动联通<a>元素。

```
print(driver.find_element_by_xpath("//div[@class='detail']/a").text)
print(driver.find_element_by_link_text("移动联通").text)
print(driver.find_element_by_partial_link_text("移动").text)
print(driver.find_element_by_partial_link_text("动联").text)
```

通过上述几种方法都能找到，结果：

```
移动联通
移动联通
移动联通
移动联通
```

但是，使用 driver.find_element_by_link_text("移动")是找不到的，因为这个函数要求文本要完全匹配。

## 5.3.9 使用 class 值查找元素

### 1. 查找单一类名的元素

selenium 可以使用元素的 class 值查找元素。

（1）find_element_by_class_name(value)函数：查找第一个 class=value 的元素，如果找到就返回该元素的 WebElement 对象，如果找不到就抛出异常。

（2）find_elements_by_class_name(value)函数：查找所有 class=value 的元素，如果找到就返回 WebElement 列表，如果找不到就返回空列表。

**例 5-3-20**：查找网页中 class="pl"的所有元素。

```
elems=driver.find_elements_by_class_name("pl")
for elem in elems:
    print(elem.text)
```

结果：

```
品牌
生产日期
价格
```

显然，也可以通过下列方式查找元素：

```
elems=driver.find_elements_by_xpath("//*[@class='pl']")
```

```
elems=driver.find_elements_by_css_selector("*[class='pl']")
```

### 2. 查找复合类名的元素

值得注意的是,在网页中,有些元素有复合的类名称。例如,修改网页的日期部分为:

```
<div class="date">
<span class="pl date">生产日期</span>:<span name="date">2016-12-01</span>
</div>
```

那么<span class="pl date">中就有复合的类名称,一个是"pl",另外一个是"date"。这种类型的元素不能使用 find_element_by_class_name()函数查找,但是可以使用 XPath 或者 CSS 语法的函数查找。例如,使用:

```
driver.find_element_by_xpath("//span[@class='pl date']")
```

或者

```
driver.find_element_by_css_selector("span[class='pl date']")
```

都可以查找。

 使用 selenium 实现用户登录

5-4-A  5-4-B
知识讲解  操作演练

**任务目标**

selenium 查找的 HTML 元素是一个 WebElement 对象,这个对象不但可以获取元素的属性值,还能完成一些键盘输入动作与鼠标单击动作。在本节中,将通过实现一个用户登录并爬取数据的程序来演示这个过程。

#### 5.4.1 创建用户登录网站

##### 1. 创建网站模板

很多网站都需要用户登录后才能访问到其他网页。在 templates 文件夹中创建一个名为 login.html 的用户登录网站模板文件,在启动时显示登录页面,用户输入名称(假定是"xxx")及密码(假定是"123")后就完成登录,然后跳转到"/show"页面显示手机记录。login.html 文件内容如下:

```
<body>
<form id="frm" action="/" method="post">
<div>用户<input type="text" name="user" id="user"></div>
<div>密码<input type="password" name="pwd" id="pwd"></div>
<div><input type="submit" name="login" id="login" vaule='登录'></div>
</form>
</body>
```

##### 2. 创建网站服务器程序

服务器程序 server.py 首先提交一个 login.html 网页,用户输入名称与密码后提交,服务器获取用户名称与密码后判断其是否正确,如果不正确就继续显示登录页面,如果正确就跳转到"/show"页面显示手机记录。服务器程序 server.py 如下:

```
import flask
app=flask.Flask(__name__)
```

```python
@app.route("/",methods=["GET","POST"])
def login():
    user=flask.request.values.get("user") if "user" in flask.request.values else ""
    pwd=flask.request.values.get("pwd") if "pwd" in flask.request.values else ""
    if user=="xxx" and pwd=="123":
        return flask.redirect("/show")
    else:
        return flask.render_template("login.html")

@app.route("/show",methods=["GET","POST"])
def show():
    s="<table border='1'>"
    s=s+"<tr><td>品牌</td><td>型号</td><td>价格</td></tr>"
    s=s+"<tr><td>华为</td><td>P9</td><td>3800</td></tr>"
    s=s+"<tr><td>华为</td><td>P10</td><td>4200</td></tr>"
    s=s+"<tr><td>苹果</td><td>iPhone6</td><td>5800</td></tr>"
    s=s+"</table><p>"
    s=s+"<a href='/'>退出</a>"
    return s

app.run()
```

其中，语句 return flask.redirect("/show")用于跳转到"/show"页面。

### 3. 使用浏览器访问

运行服务器程序后，使用浏览器访问"http://127.0.0.1:5000"，可以看到登录页面，如图 5-4-1 所示。用户登录成功后就可以看到手机的记录页面，如图 5-4-2 所示。

图 5-4-1 登录页面

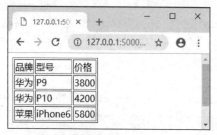
图 5-4-2 登录成功后的页面

显然，如果要爬取这个 Web 网站的所有手机记录，就必须模拟用户登录过程，使得访问到"/show"页面。

### 5.4.2 使用元素动作

selenium 查找 HTML 元素时会返回一个 WebElement 对象，这个对象的功能十分强大，不但可以使用它获取元素的属性值，还可以使用它完成键盘输入动作与鼠标单击动作。

#### 1. 键盘输入动作

有些元素（如<input type="text">文本框）是允许用户输入文字的，WebElement 对象

element 可以模拟用户的键盘输入动作，主要动作有：

（1）clear()函数用于模拟清除 element 元素中的所有文字；

（2）send_keys(string)函数用于模拟键盘在元素中输入字符串 string。

其中，send_keys()函数不但可以模拟输入一般的文字，还可以模拟输入 Enter、BackSpace 等键盘动作。selenium 提供了一个 Keys 类，其中提供了很多常用的不可见的特殊按键，如 Keys.BACKSPACE（BackSpace 键）和 Keys.ENTER（Enter 键）。

编写下面的 spider.py 程序：

```
from selenium import webdriver
import time
try:
    driver = webdriver.Chrome()
    driver.get("http://127.0.0.1:5000")
    user = driver.find_element_by_name("user")
    pwd = driver.find_element_by_name("pwd")
    time.sleep(0.5)
    user.send_keys("xxx")
    time.sleep(0.5)
    pwd.send_keys("123")
except Exception as err:
    print(err)
input("Strike any key to finish...")
driver.close()
```

这个程序使用 driver = webdriver.Chrome()创建 Chrome 浏览器，因此 Chrome 浏览器是可见的。运行该服务器程序，可以清楚地看到文本框的自动输入动作，如图 5-4-3 所示。

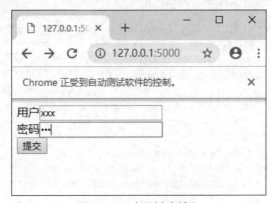

图 5-4-3　实现键盘输入

### 2. 鼠标单击动作

很多 HTML 元素都有鼠标单击动作，例如，<input type="submit">提交按钮被单击后就提交表单。WebElement 对象使用 click()函数实现鼠标单击，例如：

```
driver.find_element_by_xpath("input[@type='submit']").click()
```

### 5.4.3　编写爬虫程序

使用 selenium 模拟浏览器的访问过程与用户的登录过程，然后爬取手机的记录，程序过程如下：

（1）创建一个浏览器对象 driver，使用 driver 对象模拟浏览器。

（2）访问 "http://127.0.0.1:5000"，获取<input type="text" name="user">与<input type="password" name="pwd">元素对象，模拟用户通过键盘输入名称 "xxx" 与密码 "123"。

（3）获取<input type="submit" name="login">按钮对象，执行 click()单击动作提交表单。

（4）服务器接收提交的 user 与 pwd 数据，判断是否登录成功，如果登录成功就跳转到 "/show" 页面显示手机记录。

（5）运行爬虫程序爬取手机记录。

编写爬虫程序 spider.py，如下：

```python
from selenium import webdriver
from selenium.webdriver.chrome.options import Options
import time

def login():
    print(driver.current_url)
    user = driver.find_element_by_name("user")
    pwd = driver.find_element_by_name("pwd")
    login = driver.find_element_by_name("login")
    user.send_keys("xxx")
    pwd.send_keys("123")
    login.click()
    time.sleep(0.5)

def spider():
    print(driver.current_url)
    trs=driver.find_elements_by_tag_name("tr")
    for i in range(1,len(trs)):
        tds=trs[i].find_elements_by_tag_name("td")
        if len(tds)==3:
            mark = tds[0].text
            model=tds[1].text
            price=tds[2].text
            print("%-16s%-16s%-16s" %(mark,model,price))

chrome_options = Options()
chrome_options.add_argument('--headless')
chrome_options.add_argument('--disable-gpu')
driver = webdriver.Chrome(chrome_options=chrome_options)
try:
    driver.get("http://127.0.0.1:5000")
    login()
    spider()
except Exception as err:
    print(err)
driver.close()
```

程序先访问 "http://127.0.0.1:5000"，执行 login()登录函数，通过 user.send_keys("xxx")与 pwd.send_keys("123")向 user 与 pwd 元素发送文本字符串 "xxx" 与 "123"，这个过程与用户从键盘输入一样。

login.click()可模拟 login 按钮单击的动作，执行该命令即完成提交，提交后网页跳转到

"http://127.0.0.1:5000/show",使用 time.sleep(0.5)模拟网络延迟,等待网页稳定后执行 spider()函数爬取网页数据。

下面的执行结果表明程序成功模拟了用户的登录,并爬取到了只有登录成功后才能浏览到的数据:

```
http://127.0.0.1:5000/
http://127.0.0.1:5000/show
华为          P9              3800
华为          P10             4200
苹果          iPhone6         5800
```

### 5.4.4 执行 JavaScript 程序

#### 1. 无参数的 JavaScript 程序

实际上,这个登录过程也可以用 JavaScript 程序实现,使用 document 的 getElementById()方法找到 user 与 pwd 的对象,分别设置它们的值为 "xxx" 与 "123",再找到 login 对象并调用 click()函数完成登录。

selenium 使用 execute_script(js)函数执行 JavaScript 程序,其中 js 是要执行的 JavaScript 程序代码。因此 selenium 使用 JavaScript 程序实现登录的程序如下:

```
from selenium import webdriver
try:
    driver = webdriver.Chrome()
    driver.get("http://127.0.0.1:5000")
    js="document.getElementById('user').value='xxx';"
    js =js+ "document.getElementById('pwd').value='123';"
    js =js+ "document.getElementById('login').click();"
    driver.execute_script(js)
except Exception as err:
    print(err)
input("Strike any key to finish...")
driver.close()
```

#### 2. 有参数的 JavaScript 程序

selenium 在执行 execute_script(js)时还可以使用 selenium 的 WebElement 对象参数代替 JavaScript 中的 DOM 对象参数,此时在 js 中使用 arguments[n]形式参数,其中 n=0,1,2,…,表示参数的顺序。

例如,使用 selenium 找到 user 对象并设置它的值为 "xxx" 的语句如下:

```
user=driver.find_element_by_id("user")
driver.execute_script("arguments[0].value='xxx'",user)
```

这里的 arguments[0]是参数,执行时被 WebElement 对象 user 代替,效果等同于下面的语句:

```
driver.execute_script("document.getElementById('user').value='xxx'")
```

为 user 与 pwd 设置值可以使用下面的程序:

```
user=driver.find_element_by_id("user")
pwd = driver.find_element_by_id("pwd")
driver.execute_script("arguments[0].value='xxx';",user)
```

```
driver.execute_script("arguments[0].value='123';",pwd)
```

当然也可以将后面两条语句合并在一起执行。下面程序的效果与上述程序的效果一致。

```
user=driver.find_element_by_id("user")
pwd = driver.find_element_by_id("pwd")
driver.execute_script("arguments[0].value='xxx';arguments[1].value='123';",user,pwd)
```

执行 execute_script()时有两个参数——arguments[0]与 arguments[1],它们将被实际的 user 与 pwd 代替。

根据这些规则,爬虫程序的 login()函数可以重新编写如下:

```
def login():
    print(driver.current_url)
    user = driver.find_element_by_name("user")
    pwd = driver.find_element_by_name("pwd")
    login = driver.find_element_by_name("login")
    driver.execute_script("arguments[0].value='xxx';",user)
    driver.execute_script("arguments[0].value='123';",pwd)
    driver.execute_script("arguments[0].click();", login)
    time.sleep(0.5)
```

该函数的效果与之前的效果完全一样。

## 5.5 使用 selenium 爬取 Ajax 网页数据

5-5-A 知识讲解　5-5-B 操作演练

### 任务目标

现在的网页中大量使用 Ajax 技术,通过 JavaScript 程序使客户端向服务器发出请求,服务器返回数据给客户端,客户端再把数据展现出来,这样做可以减少网页的闪动,让用户有更好的体验。在本节中先设计一个类似的网页,然后使用 selenium 编写爬虫程序来爬取网页的数据。

### 5.5.1 创建 Ajax 网站

**1. 创建网页文件**

在 templates 文件夹中创建网页文件 phone.html,它的内容如下:

```
<script>
    function init()
    {
      var marks=new Array("华为","苹果","三星");
      var selm=document.getElementById("marks");
      for(var i=0;i<marks.length;i++)
      {
          selm.options.add(new Option(marks[i],marks[i]));
      }
      selm.selectedIndex=0;
      display();
    }
```

```
        function display()
        {
            try
            {
                var http=new XMLHttpRequest();
                var selm=document.getElementById("marks");
                var m=selm.options[selm.selectedIndex].text;
                http.open("get","/phones?mark="+m,false);
                http.send(null);
                msg=http.responseText;
                obj=eval("("+msg+")");
                s="<table width='200' border='1'><tr><td>型号</td><td>价格</td></tr>"
                for(var i=0;i<obj.phones.length;i++)
                {
                    s=s+"<tr><td>"+obj.phones[i].model+"</td><td>"+obj.phones[i].price+"</td></tr>";
                }
                s=s+"</table>";
                document.getElementById("phones").innerHTML=s;
            }
            catch(e) { alert(e); }
        }
</script>
<body onload="init()">
<div>选择品牌<select id="marks" onchange="display()"></select></div>
<div id="phones"></div>
</body>
```

程序说明如下。

(1) 网页框架。

这个网页的主体框架很简单,只有两个<div>元素,一个包含<select>元素,另一个用于显示信息,在执行 JavaScript 程序之前内容都是空的。

(2) init()函数。

当网页被加载时执行 init()函数,为<select>增加 3 个手机品牌,即"华为""苹果""三星"。

(3) display()函数。

只要用户选择其中一个品牌就触发<select>的 onchange 事件,从而执行 display()函数。该函数通过 Ajax 技术把选择的手机品牌通过 http.open("get","/phones?mark="+m,false)语句发送给服务器,服务器收到该手机品牌后将这个品牌的手机信息 http.responseText 返回给这个网页。返回的数据是 JSON 格式的字符串,经 eval()转换为 JavaScript 对象 phones 后,将生成一张表格的 HTML 代码于<div id="phones">…</div>中显示。

### 2. 创建服务器程序

要创建服务器程序 server.py,首先要提交一个 phone.html 网页,然后响应"/phones?mark=…"的请求,根据 mark 的值确定品牌,返回该品牌下的手机记录,返回的记录采用 JSON 格式表示。服务器程序 server.py 如下:

```python
import flask
import json
app=flask.Flask(__name__)
@app.route("/")
def index():
    return flask.render_template("phone.html")

@app.route("/phones")
def getPhones():
    mark=flask.request.values.get("mark")
    phones=[]
    if mark=="华为":
        phones.append({"model":"P9","mark":"华为","price":3800})
        phones.append({"model": "P10", "mark": "华为", "price": 4000})
    elif mark=="苹果":
        phones.append({"model":"iPhone5","mark":"苹果","price":5800})
        phones.append({"model":"iPhone6","mark":"苹果","price":6800})
    elif mark=="三星":
        phones.append({"model":"Galaxy A9","price":2800})
    s=json.dumps({"phones":phones})
    return s
if __name__=="__main__":
    app.run()
```

### 3. 使用浏览器访问

运行服务器程序，使用浏览器访问 "http://127.0.0.1:5000"，结果如图 5-5-1 所示。用户选择另外一个品牌后触发 phone.html 中<select>的 onchange 事件，再次使用 Ajax 获取该品牌的手机记录进行显示。

图 5-5-1　手机网站

## 5.5.2　理解 selenium 爬虫程序

### 1. 普通爬虫程序

编写如下爬虫程序 spider.py 以获取网页的 HTML 代码。

```
import urllib.request
resp=urllib.request.urlopen("http://127.0.0.1:5000")
html=resp.read().decode()
print(html)
```

执行该程序，结果如下：

```
<script>
    function init()
    {
        var marks=new Array("华为","苹果","三星");
        var selm=document.getElementById("marks");
        for(var i=0;i<marks.length;i++)
        {
            selm.options.add(new Option(marks[i],marks[i]));
        }
        selm.selectedIndex=0;
        display();
    }

    function display()
    {
        try
        {
            var http=new XMLHttpRequest();
            var selm=document.getElementById("marks");
            var m=selm.options[selm.selectedIndex].text;
            http.open("get","/phones?mark="+m,false);
            http.send(null);
            msg=http.responseText;
            obj=eval("("+msg+")");
            s="<table width='200' border='1'><tr><td>型号</td><td>价格</td></tr>"
            for(var i=0;i<obj.phones.length;i++)
            {
                s=s+"<tr><td>"+obj.phones[i].model+"</td><td>"+obj.phones[i].price+"</td></tr>";
            }
            s=s+"</table>";
            document.getElementById("phones").innerHTML=s;
        }
        catch(e) { alert(e); }
    }
</script>
<body onload="init()">
<div>选择品牌<select id="marks" onchange="display()"></select></div>
<div id="phones"></div>
</body>
```

显然，如果要爬取这个 Web 网站的所有手机记录，采用简单的爬虫程序是得不到的，因为在网页的 HTML 代码中根本看不出任何手机的记录信息。

实际上，在使用 BeautifulSoup 解析时，BeautifulSoup 根据网页的 HTML 代码构建了 DOM 网页树；在使用 scrapy 的 Selector 解析时，Selector 也根据网页的 HTML 代码构建了 DOM 网页树。无论是 BeautifulSoup 还是 Selector，都只能根据静态的 HTML 代码构建 DOM 网页树，它们都不能使用 JavaScript 程序构建 DOM 网页树。

2. selenium 爬虫程序

selenium 是一个浏览器，浏览一个网页后会在内部构建一棵由 HTML 元素组成的 DOM

网页树。编写如下 spider.py 程序以获取网页的 HTML 代码。

```python
from selenium import webdriver
from selenium.webdriver.chrome.options import Options
chrome_options = Options()
chrome_options.add_argument('--headless')
chrome_options.add_argument('--disable-gpu')
driver = webdriver.Chrome(chrome_options=chrome_options)
driver.maximize_window()
driver.get("http://127.0.0.1:5000")
print(driver.page_source)
driver.close()
```

执行该程序，结果如下：

```
<html xmlns="http://www.w3.org/1999/xhtml"><head><script>
    function init()
    {
        var marks=new Array("华为","苹果","三星");
        var selm=document.getElementById("marks");
        for(var i=0;i&lt;marks.length;i++)
        {
            selm.options.add(new Option(marks[i],marks[i]));
        }
        selm.selectedIndex=0;
        display();
    }

    function display()
    {
        try
        {
            var http=new XMLHttpRequest();
            var selm=document.getElementById("marks");
            var m=selm.options[selm.selectedIndex].text;
            http.open("get","/phones?mark="+m,false);
            http.send(null);
            msg=http.responseText;
            obj=eval("("+msg+")");
            s="&lt;table width='200' border='1'&gt;&lt;tr&gt;&lt;td&gt;型号&lt;/td&gt;&lt;td&gt;价格&lt;/td&gt;&lt;/tr&gt;"
            for(var i=0;i&lt;obj.phones.length;i++)
            {
                s=s+"&lt;tr&gt;&lt;td&gt;"+obj.phones[i].model+"&lt;/td&gt;&lt;td&gt;"+obj.phones[i].price+"&lt;/td&gt;&lt;/tr&gt;";
            }
            s=s+"&lt;/table&gt;";
            document.getElementById("phones").innerHTML=s;
        }
        catch(e) { alert(e); }
    }
</script>
</head><body onload="init()">
    <div>选择品牌<select id="marks" onchange="display()"><option value="华为">
```

```
华为</option><option value="苹果">苹果</option><option value="三星">三星
</option></select></div>
    <div id="phones"><table width="200" border="1"><tbody><tr><td>型号</td>
<td>价格</td></tr><tr><td>P9</td><td>3800</td></tr><tr><td>P10</td><td>
4000</td></tr></tbody></table></div>
    </body></html>
```

由此可见,这棵树的元素包括静态 HTML 的元素,也包括 JavaScript 程序执行后产生的元素,selenium 会根据这棵树去查找各个元素。

### 5.5.3 编写爬虫程序

首先使用 selenium 模拟浏览器去访问网页,然后模拟用户选择<select>中各个手机品牌的过程实现换页,就可以逐页爬取所有的手机数据了。爬虫程序执行过程如下。

(1)创建一个浏览器对象 driver,使用 driver 对象模拟浏览器。
(2)访问 "http://127.0.0.1:5000",爬取第一个页面的手机数据。
(3)从第一个页面中获取<select>中所有的 options。
(4)循环 options 中的每个 option,并模拟这个 option 的单击动作,触发 onchange 事件,调用 display()函数,爬取每个页面的手机数据。

根据这个规则编写爬虫程序 spider.py,如下:

```python
from selenium import webdriver
from selenium.webdriver.chrome.options import Options
import time

def spider():
    trs=driver.find_elements_by_tag_name("tr")
    for i in range(1,len(trs)):
        tds=trs[i].find_elements_by_tag_name("td")
        model=tds[0].text
        price=tds[1].text
        print("%-16s%-16s" %(model,price))

chrome_options = Options()
chrome_options.add_argument('--headless')
chrome_options.add_argument('--disable-gpu')
driver = webdriver.Chrome(chrome_options=chrome_options)
driver.maximize_window()
driver.get("http://127.0.0.1:5000")
select = driver.find_element_by_id("marks")
options = select.find_elements_by_tag_name("option")
for option in options:
    option.click()
    spider()
driver.close()
```

程序说明如下。

(1)创建 Chrome 对象:

```
driver = webdriver.Chrome(chrome_options=chrome_options)
```

然后访问网站的第一页:

```
driver.get("http://127.0.0.1:5000")
```

（2）spider()函数负责爬取当前页面的所有手机记录，其中：
```
trs=driver.find_elements_by_tag_name("tr")
```
用于获取所有的<tr>元素。网页中有很多<tr>元素，可通过循环去获取这些<tr>元素。程序跳过第一个<tr>的表头，从第二个<tr>开始是手机的记录。

再通过：
```
tds=trs[i].find_elements_by_tag_name("td")
```
获取<tr>下面的所有<td>元素。每个<tr>下面都有两个<td>元素，第一个包含手机的型号，第二个包含手机的价格：
```
model=tds[0].text
price=tds[1].text
```
这样就爬取到了各个手机记录。

（3）程序通过：
```
select=driver.find_element_by_id("marks")
```
获取网页中的<select>元素，再通过
```
options=select.find_elements_by_tag_name("option")
```
获取该元素下所有的<option>元素。

（4）循环每个 options 元素：
```
for option in options:
    option.click()
    spider()
```
然后为每个元素调用 option.click()函数。每次执行 option.click()都是一个模拟用户单击该<option>的动作，它会触发<select>的 onchange 事件，从而执行 display()函数，用 Ajax 从服务器获取该手机品牌的手机记录，再次调用 spider()函数就可以爬取这些手机记录。

注意，每次执行 option.click()后，网页中的<select>元素整体是没有改变的，因此可以使用循环的方法连续调用 option.click()。

### 5.5.4 执行爬虫程序

执行爬虫程序，爬取到了所有的手机记录，如下：
```
P9            3800
P10           4000
iPhone5       5800
iPhone6       6800
Galaxy A9     2800
```
这些手机记录也就是各个页面的手机记录的总和，说明爬虫程序爬取成功。

## 5.6 使用 selenium 等待 HTML 元素

5-6-A 知识讲解　5-6-B 操作演练

**任务目标**

在浏览器加载网页的过程中，网页的有些元素会出现延迟的现象，在 HTML 元素还没有准备好的情况下去操作这个 HTML 元素必然会出现错误，这时 selenium 需要等待延迟的 HTML 元素。在本节中，我们主要学习使用 selenium 等待延迟的 HTML 元素并最终爬取元素的数据。

## 5.6.1 创建延迟模拟网站

### 1. 创建网站模板

在 templates 文件夹中创建模板文件 phone.html，这个文件使用 Ajax 从服务器获取手机的品牌数据并将其放在<select>中。注意，<select>中的<option>开始时是不存在的，只有获取数据才产生。模板文件 phone.html 如下：

```
<script>
    function loadMarks()
    {
       var http=new XMLHttpRequest();
       http.onreadystatechange=function()
       {
          if (http.readyState==4 && http.status==200)
          {
             var xmark=document.getElementById("xmark");
             var xcolor=document.getElementById("xcolor");
             marks=eval("("+http.responseText+")");
             for(var i=0;i<marks.length;i++)
             xmark.options.add(new Option(marks[i],marks[i]));
             document.getElementById("submit").disabled=false;
             document.getElementById("msg").innerHTML="品牌";
          }
       };
       http.open("get","/marks",true);
       http.send(null);
    }
    loadMarks();
</script>
<body>
<form name="frm" action="/">
 <div><span id="msg"></span><select id="xmark" ></select></div>
<input type="submit" value="提交" id="submit" disabled="true">
</form>
</body>
```

### 2. 创建网站服务器程序

网站服务器在访问地址"/"时首先提交 phone.html 网页，然后网页中根据 JavaScript 程序执行 loadMarks()函数，再次访问服务器"/phones"时发送手机的相关数据，数据按 JSON 字符串格式发送。为了模拟延迟过程，使用 time.sleep(1)延迟 1s 后发送数据。服务器程序 server.py 如下：

```
import flask
import json
import time
app=flask.Flask(__name__)
@app.route("/")
def index():
    return flask.render_template("phone.html")
```

```python
@app.route("/marks")
def loadMarks():
    time.sleep(1)
    marks=["华为","苹果","三星"]
    return json.dumps(marks)
app.run()
```

运行服务器程序，并使用浏览器访问"http://127.0.0.1:5000"，延迟 1s 后出现图 5-6-1 所示的网页。

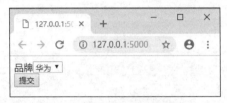

图 5-6-1　延迟 1s 后出现的网页

### 5.6.2　编写爬虫程序

编写爬虫程序 spider.py 去爬取手机的所有品牌与颜色数据，并选择其中的一个品牌与颜色数据进行提交。爬虫程序 spider.py 编写如下：

```python
from selenium import webdriver
import time
driver = webdriver.Chrome()
driver.get("http://127.0.0.1:5000")
marks=driver.find_elements_by_xpath("//select/option")
print("品牌数量:",len(marks))
for mark in marks:
    print(mark.text)
form=driver.find_element_by_xpath("//form")
print(form.get_attribute("innerHTML").strip())
time.sleep(5)
driver.close()
```

执行该程序，结果如下：

```
品牌数量: 0
<div>品牌<select id="xmark"></select></div>
<input type="submit" value="提交" id="submit" disabled="true">
```

由此可见，这个爬虫程序没有爬取到手机品牌与颜色的数据，原因是服务器有延迟，这些数据还没有在网页中生成。

### 5.6.3　selenium 强制等待

selenium 使用 time.sleep(seconds)来实现强制等待 seconds 秒，这是最简单、粗暴的方式，不管当前操作是否完成，是否可以进行下一步操作，都必须等待 seconds 秒。

这种方式的优点是简单，缺点是不能准确把握需要等待的时间（有时操作还未完成，等待就结束了，导致报错；有时操作已经完成了，但等待时间还没有到，浪费时间），如果在实际中大量使用，会浪费不必要的等待时间，影响测试用例的执行效率。

改进爬虫程序 spider.py，让它在加载网页后强制等待 1.5s，程序如下：

```
from selenium import webdriver
import time
driver = webdriver.Chrome()
driver.get("http://127.0.0.1:5000")
#设置强制等待1.5s
time.sleep(1.5)
marks=driver.find_elements_by_xpath("//select/option")
print("品牌数量:",len(marks))
for mark in marks:
    print(mark.text)
form=driver.find_element_by_xpath("//form")
print(form.get_attribute("innerHTML").strip())
time.sleep(5)
driver.close()
```

执行结果：

品牌数量: 3
华为
苹果
三星
品牌<select id="xmark"><option value="华为">华为</option><option value="苹果">苹果</option><option value="三星">三星</option></select>
<input type="submit" value="提交" id="submit">

由此可见，等待1.5s后程序从服务器获取了手机品牌数据并创建了<select>中的各个<option>元素，因此程序爬取到手机品牌数据。但是，如果设置的强制等待时间不够长，那么还是爬取不到需要的数据。

### 5.6.4 selenium 隐式等待

selenium 使用 implicitly_wait(seconds)设置隐性等待指定的秒数，即网页在加载时最长等待 seconds 秒，例如，爬虫程序在访问网页时设置隐性等待 1.5s。spider.py 程序如下：

```
from selenium import webdriver
import time
driver = webdriver.Chrome()
#设置隐式等待1.5s
driver.implicitly_wait(1.5)
driver.get("http://127.0.0.1:5000")
marks=driver.find_elements_by_xpath("//select/option")
print("品牌数量:",len(marks))
for mark in marks:
    print(mark.text)
form=driver.find_element_by_xpath("//form")
print(form.get_attribute("innerHTML").strip())
time.sleep(5)
driver.close()
```

执行结果：

品牌数量: 3
华为

```
苹果
三星
<div>品牌<select id="xmark"><option value="华为">华为</option><option value="苹果">苹果</option><option value="三星">三星</option></select></div>
<input type="submit" value="提交" id="submit">
```

由此可见，经过等待后程序从服务器获取了手机品牌数据并创建了<select>中的各个<option>元素，因此程序爬取到了手机品牌数据。同样，如果设置的隐性等待时间不够长，那么还是爬取不到需要的数据。

### 5.6.5 selenium 循环等待与显式等待

#### 1. 循环等待

实际上这个爬虫程序能否爬取到数据的关键是<select>中是否已经出现了<option>元素，我们可以设置一个循环来判断是否出现了<option>元素。程序 spider.py 修改如下：

```python
from selenium import webdriver
import time
driver = webdriver.Chrome()
try:
    driver.get("http://127.0.0.1:5000")
    waitTime=0
    while waitTime<10:
        marks = driver.find_elements_by_xpath("//select/option")
        if len(marks)>0:
            break
        time.sleep(0.5)
        waitTime+=0.5
    if waitTime>=10:
        raise Exception("Waiting time out")
    marks=driver.find_elements_by_xpath("//select/option")
    print("品牌数量:",len(marks))
    for mark in marks:
        print(mark.text)
    form=driver.find_element_by_xpath("//form")
    print(form.get_attribute("innerHTML").strip())
except Exception as err:
    print(err)
time.sleep(5)
driver.close()
```

这个程序使用 waitTime 变量来构造循环，它最长等待 10s，每间隔 0.5s 就检查一次<select>中是否出现<option>元素，如果出现<option>元素则退出等待循环，否则继续等待，直到<option>元素出现为止，如果 10s 后还没有出现<option>元素就抛出异常。

执行这个爬虫程序，结果如下：

```
品牌数量：3
华为
苹果
三星
```

```html
    <div>品牌<select id="xmark"><option value="华为">华为</option><option value="苹果">苹果</option><option value="三星">三星</option></select></div>
    <input type="submit" value="提交" id="submit">
```

显然这种循环等待的方式比强制等待的方式好，不会出现<option>元素出现后还继续等待的情况。

### 2. 显式等待

selenium 的显式等待与循环等待有点类似，它是专门等待指定元素的。selenium 使用 WebDriverWait 类来实现显式等待，在实现显式等待之前先引入 WebDriverWait、EC 以及 By 等类：

```python
from selenium.webdriver.support.wait import WebDriverWait
from selenium.webdriver.support import expected_conditions as EC
from selenium.webdriver.common.by import By
```

然后构造一个定位元素的 locator 对象。例如，通过 XPath 方法定位<select>中的<option>元素：

```python
locator=(By.XPATH,"//select/option")
```

最后使用 WebDriverWait 构造一个实例，调用 until()方法：

```python
WebDriverWait(driver,10, 0.5).until(EC.presence_of_element_located(locator))
```

这条语句的含义是等待 locator 指定的元素出现，最长等待 10s，每间隔 0.5s 就检查一次。如果在 10s 内出现了该元素则结束等待，否则继续等待。如果超过 10s 还没有等到 locator 指定的元素就抛出异常。

使用显式等待的爬虫程序 spider.py 如下：

```python
from selenium import webdriver
import time
from selenium.webdriver.support.wait import WebDriverWait
from selenium.webdriver.support import expected_conditions as EC
from selenium.webdriver.common.by import By
driver = webdriver.Chrome()
try:
    driver.get("http://127.0.0.1:5000")
    locator = (By.XPATH, "//select/option")
    WebDriverWait(driver, 10,0.5).until(EC.presence_of_element_located(locator))
    marks=driver.find_elements_by_xpath("//select/option")
    print("品牌数量:",len(marks))
    for mark in marks:
        print(mark.text)
    form=driver.find_element_by_xpath("//form")
    print(form.get_attribute("innerHTML").strip())
except Exception as err:
    print(err)
time.sleep(5)
driver.close()
```

执行结果：

```
品牌数量: 3
```

```
华为
苹果
三星
<div>品牌<select id="xmark"><option value="华为">华为</option><option
value="苹果">苹果</option><option value="三星">三星</option></select></div>
<input type="submit" value="提交" id="submit">
```

显然，程序等到了<option>元素的出现，爬取到了手机品牌数据。显式等待的优点就是等待时间判断准确，不会浪费多余的等待时间，在实际中使用可以提高执行效率。

### 5.6.6 selenium 显式等待形式

显式等待有很多种形式，读者可以查看 selenium 的文档说明，下面是一些常用的形式。

#### 1. EC.presence_of_element_located(locator)

这种形式是等待 locator 指定的元素出现，也就是等待 HTML 文档中建立起该元素。

#### 2. EC.visibility_of_element_located(locator)

这种形式是等待 locator 指定的元素可见。注意，元素出现时未见得可见。例如：

```
<select id="xmark" style="display:none">...</select>
```

虽然<select>元素出现了，但是其不可见。

#### 3. EC.element_to_be_clickable(locator)

这种形式是等待 locator 指定的元素可以被单击。例如，在爬虫程序中等待<input type="submit">可以被单击：

```
locator = (By.XPATH, "//input[@type='submit']")
WebDriverWait(driver, 10,0.5).until(EC.element_to_be_clickable(locator))
```

或者等待<option>是否可以被单击：

```
locator = (By.XPATH, "//select/option")
WebDriverWait(driver, 10,0.5).until(EC.element_to_be_clickable(locator))
```

使用这两种方法都可以爬取到手机品牌数据。
但是注意，使用：

```
locator = (By.XPATH, "//select")
WebDriverWait(driver, 10,0.5).until(EC.element_to_be_clickable(locator))
```

表示等待<select>是否可以被单击，这个元素在没有<option>时也是可以被单击的，因此用这种等待方式是爬取不到手机的品牌数据的。

#### 4. EC.element_located_to_be_selected(locator)

这种形式是等待 locator 指定的元素可以被选择，可以被选择的元素一般是<select>中的<option>、<input type="checkbox">以及<input type="radio">等元素。例如，在爬虫程序中使用：

```
locator = (By.XPATH, "//select/option")
WebDriverWait(driver, 10,0.5).until(EC.element_located_to_be_selected
(locator))
```

同样能爬取到手机的品牌数据。
但是使用下列语句是不行的：

```
locator = (By.XPATH, "//input[@type='submit']")
```

```
    WebDriverWait(driver, 10,0.5).until(EC.element_located_to_be_selected
(locator))
```

因为<input type='submit'>不可以被选择,这种元素不是可选择元素。

### 5. EC.text_to_be_present_in_element(locator,text)

这种形式是等待 locator 指定元素的文本中包含指定的 text 文本。例如,在爬虫程序中使用下列语句等待:

```
    locator = (By.ID, "msg")
    WebDriverWait(driver, 10,0.5).until(EC.text_to_be_present_in_element
(locator,"品"))
```

即等待<span id="msg">…</span>元素中的文本包含"品"字,由于在<option>出现后设置文本为"品牌",因此爬虫程序可以爬取到手机品牌数据。

总而言之,在使用 selenium 爬取动态网页数据时,必须使用恰当的等待方式去等待动态的网页元素生成完毕后才能操作这些元素。

## 5.7 综合项目 爬取模拟商城网站数据

任务目标

在本项目中,首先在本地建立一个模拟商城网站,用多个页面展示手机的数据,然后使用 selenium 编写爬虫程序爬取商城中手机的数据与图像,并把数据存储到数据库中。

5-7-A 知识讲解　　5-7-B 操作演练

### 5.7.1 创建模拟商城网站

#### 1. 创建网站模板

在 templates 文件夹中创建模板文件 phone.html,内容如下:

```html
<style>
  .phone
  {
    display:inline-block;
    padding:10px 10px 10px 10px;
    border:solid 0px;
    width:200px;
    height:300px;
    vertical-align: text-top;
    text-align:left
  }
</style>

<script>
    function firstPage()
    {
       window.location.href="/?pageIndex=1";
    }

    function prevPage()
```

```
        {
            var pageIndex=parseInt(document.getElementById("pageIndex").value);
            pageIndex--;
            if (pageIndex>=1)
            {
                window.location.href="/?pageIndex="+pageIndex;
            }
        }

        function nextPage()
        {
            var pageIndex=parseInt(document.getElementById("pageIndex").value);
            var pageCount=parseInt(document.getElementById("pageCount").value);
            pageIndex++;
            if (pageIndex<=pageCount)
            {
                window.location.href="/?pageIndex="+pageIndex;
            }
        }

        function lastPage()
        {
            var pageCount=parseInt(document.getElementById("pageCount").value);
            window.location.href="/?pageIndex="+pageCount;
        }
</script>
<body>
<div style="text-align:center">
{% for p in phones %}
{% if loop.index0 is divisibleby(rowItemCount) %}
    <div style="width:{{rowItemCount*250}}px;text-align:left;display:inline-block">
{% endif %}
<div class="phone"  >
<div>{{p.no}}</div>
<div><img src="images/{{p.image}}" width="100" height="100"></div>
<p></p>
<div>{{p.mark}}</div>
<div>价格:¥<span style="color:red">{{p.price}}</span></div>
<div>{{p.note}}</div>
</div>
{% if (loop.index0+rowItemCount+1) is divisibleby(rowItemCount) %}
</div>
{% endif %}
{% endfor %}
{% if numbers is not divisibleby(rowItemCount) %}
</div>
{% endif %}
</div>
<div align="center" class="paging">
<input type="hidden" id="pageIndex" value="{{pageIndex}}">
<input type="hidden" id="pageCount" value="{{pageCount}}">
```

```
<input type="button" id="first" value="第一页" onclick="firstPage()">
    {% if pageIndex>1 %}
<input type="button"  value="前一页" onclick="prevPage()">
    {% else %}
<input type="button"  value="前一页" disabled="true">
    {% endif %}
    {% if pageIndex<pageCount %}
<input type="button" id="next" value="下一页" onclick="nextPage()">
    {% else %}
<input type="button"  value="下一页" disabled="true">
    {% endif %}
<input type="button" id="last" value="最后页" onclick="lastPage()">
<span>Page {{pageIndex}}/{{pageCount}}</span>
</div>
</body>
```

这个模板文件包含几个参数。

（1）phones 参数：表示所有手机的记录列表，每个列表元素都是一个字典，字典中包括手机品牌 mark、价格 price、简介 note、图像 image 等数据，使用循环{% for p in phones %}得到每部手机的数据。

（2）rowItemCount 参数：指定一行显示几部手机的信息。使用：

```
{% if loop.index0 is divisibleby(rowItemCount) %}
<div style="width:{{rowItemCount*250}}px;text-align:left;display:inline-block">
{% endif %}
```

来开启一个<div>元素以显示 rowItemCount 部手机的信息，每部手机信息显示的宽度是 250px。再使用：

```
{% if (loop.index0+rowItemCount+1) is divisibleby(rowItemCount) %}
</div>
{% endif %}
```

来建立一个</div>进行封闭。

（3）numbers 参数：表示手机的总数量。使用它测试：

```
{% if numbers is not divisibleby(rowItemCount) %}
</div>
{% endif %}
```

即当总数量不是 rowItemCount 的倍数时增加一个</div>的封闭元素。

（4）pageIndex 与 pageCount 参数：程序使用两个隐藏的<input type="hidden">元素记录这两个参数的值。

```
<input type="hidden" id="pageIndex" value="{{pageIndex}}">
<input type="hidden" id="pageCount" value="{{pageCount}}">
```

pageCount 参数表示总页数，pageIndex 参数表示当前页码，其中 pageIndex=1,2,…,pageCount。使用<input type="button">来构造翻页按钮，如"下一页"按钮：

```
{% if pageIndex<pageCount %}
<input type="button" id="next" value="下一页" onclick="nextPage()">
    {% else %}
<input type="button"  value="下一页" disabled="true">
    {% endif %}
```

当 pageIndex<pageCount 时该按钮可用，单击后执行 nextPage()函数；当 pageIndex=pageCount 时该按钮不可用。

### 2. 创建网站服务器程序

使用 phones.csv 文件中的数据构建一个有多个页面的模拟商城网站，服务器使用 pageRowCount 控制一个页面的行数，使用 rowItemCount 控制一行显示几部手机的信息，它读取 phones.csv 文件中的数据，向 phone.html 提交 phones、numbers、pageIndex、pageCount 等参数。服务器程序 server.py 编写如下：

```python
import flask
app=flask.Flask(__name__,static_folder="images")

@app.route("/")
def show():
    pageRowCount=3
    rowItemCount = 4
    if "pageIndex" in flask.request.values:
        pageIndex=int(flask.request.values.get("pageIndex"))
    else:
        pageIndex=1
    startRow=(pageIndex-1)*pageRowCount*rowItemCount
    endRow=pageIndex*pageRowCount*rowItemCount
    phones=[]
    try:
        fobj=open("phones.csv","r",encoding="utf-8")
        rows=fobj.readlines()
        count=0
        for row in rows:
            if row.strip().strip("\n")!="":
                count=count+1
        pageCount=count//(pageRowCount*rowItemCount)
        if count % (pageRowCount*rowItemCount)!=0:
            pageCount+=1
        rowIndex=0
        for i in range(1,count):
            row=rows[i]
            if rowIndex>=startRow and rowIndex<endRow:
                row=row.strip().strip("\n")
                s=row.split(",")
                phones.append({"no":s[0],"mark":s[1],"price":s[2],"note": s[3], "image": s[4]})
            rowIndex+=1
        fobj.close()
    except Exception as err:
        print(err)
    return flask.render_template("phone.html",phones=phones,rowItemCount=rowItemCount,numbers=len(phones),pageIndex=pageIndex,pageCount=pageCount)

app.run()
```

运行服务器程序，并使用浏览器访问 "http://127.0.0.1:5000"，结果如图 5-7-1 所示。

# 项目 ❺ 爬取商城网站数据

图 5-7-1 模拟商城网站

## 5.7.2 爬取网站数据并实现网页翻页

### 1. 爬取网站数据

网站的网页结构实际上非常简单，例如，手机的 HTML 代码如下：

```
<div class="phone"  >
<div>000001</div>
<div><img src="images/000001.jpg" width="100" height="100"></div>
<p></p>
<div>荣耀 9i</div>
<div>价格：¥<span style="color:red">1198.0</span></div>
<div>荣耀 9i 4GB+64GB 幻夜黑 移动联通电信 4G 全面屏手机 双卡双待</div>
</div>
```

因此，使用下列方法可以爬取手机的图像地址 src、品牌 mark、价格 price、简介 note：

```
divs = self.driver.find_elements_by_xpath("//div[@class='phone']")
for div in divs:
        src = div.find_element_by_xpath(".//div[position()=2]//img").get_attribute("src")
        src=urllib.request.urljoin(self.driver.current_url,src)
        mark = div.find_element_by_xpath(".//div[position()=3]").text
        price = div.find_element_by_xpath(".//div[position()=4]").text
        note = div.find_element_by_xpath(".//div[position()=5]").text
```

## 2. 实现网页翻页

在网页中查找"下一页"按钮,它是<div class="paging">中所有<input type="button">中的倒数第二个,找到后查看它是否有效,如果有效就单击它进入下一页,程序如下:

```
try:
    link = self.chrome.find_elements_by_xpath("//div[@class=
'paging']//input[@type='button']")[-2]
    if link.is_enabled():
        link.click()
```

其中,link.is_enabled()用于测试 link 按钮是否有效,如果有效就单击,执行 JavaScript 的 nextPage()函数跳转到下一页。

### 5.7.3 设计数据存储与图像存储

#### 1. 数据存储

使用 SQLite3 数据库 phones.db 存储数据,该数据库中有一张 phones 表,表结构如表 5-7-1 所示。

表 5-7-1 phones 表结构

| 字段 | 类型 | 说明 |
| --- | --- | --- |
| mNo | varchar(8) | 编号(关键字) |
| mName | varchar(256) | 手机名称 |
| mMark | varchar(256) | 手机品牌 |
| mPrice | varchar(64) | 手机价格 |
| mNote | varchar(1024) | 手机简介 |
| mFile | varchar(256) | 图像名称 |

下面这段程序在每次爬取数据之前都保证建立了 phones 表而且表为空:

```
self.con = sqlite3.connect("phones.db")
self.cursor = self.con.cursor()
try:
    self.cursor.execute("drop table phones")
except:
    pass
try:
    sql = "create table phones (mNo varchar(8) primary key,mMark varchar(256),mPrice varchar(64),mNote varchar(1024),mFile varchar(32))"
    self.cursor.execute(sql)
except:
    pass
```

#### 2. 图像存储

图像存储在 download 文件夹中。下面这段程序在每次爬取图像之前都保证建立了 download 文件夹,而且先删除 download 文件夹中已经存在的所有文件:

```
if not os.path.exists(MySpider.imagePath):
    os.mkdir(MySpider.imagePath)
images = os.listdir(MySpider.imagePath)
```

```
        for img in images:
            s = os.path.join(MySpider.imagePath, img)
            os.remove(s)
```

### 5.7.4 编写爬虫程序

根据前面的分析,我们使用 selenium 编写爬虫程序 spider.py,如下:

```
from selenium import webdriver
from selenium.webdriver.chrome.options import Options
import urllib.request
import threading
import sqlite3
import os
import time

class MySpider:
    #设置 Use-Agent 值,模拟浏览器
    headers = {
        "User-Agent": "Mozilla/5.0 (Windows; U; Windows NT 6.0 x64; en-US; rv:1.9pre) Gecko/2008072421 Minefield/3.0.2pre"}
    #下载图像的路径
    imagePath = "download"

    def startUp(self):
        #初始化函数
        # Initializing Chrome browser
        chrome_options = Options()
        chrome_options.add_argument('--headless')
        chrome_options.add_argument('--disable-gpu')
        self.driver = webdriver.Chrome(chrome_options=chrome_options)
        self.driver.maximize_window()

        # Initializing variables
        #初始化线程列表
        self.threads = []
        self.No = 0

        # Initializing database
        #初始化数据库,建立 phones 表
        try:
            self.con = sqlite3.connect("phones.db")
            self.cursor = self.con.cursor()
            try:
                self.cursor.execute("drop table phones")
            except:
                pass
            try:
                sql = "create table phones (mNo varchar(8) primary key,mMark varchar(256),mPrice varchar(64),mNote varchar(1024),mFile varchar(32))"
                self.cursor.execute(sql)
            except:
```

```python
            pass
        except Exception as err:
            print(err)

        #初始化图像文件夹，删除原有的图像
        try:
            if not os.path.exists(MySpider.imagePath):
                os.mkdir(MySpider.imagePath)
            images = os.listdir(MySpider.imagePath)
            for img in images:
                s = os.path.join(MySpider.imagePath, img)
                os.remove(s)
        except Exception as err:
            print(err)

    def closeUp(self):
        #结束函数，关闭数据库与浏览器
        try:
            self.con.commit()
            self.con.close()
            self.driver.close()
        except Exception as err:
            print(err);

    def insertDB(self, mNo, mMark, mPrice, mNote, mFile):
        #插入数据记录
        try:
            sql = "insert into phones (mNo,mMark,mPrice,mNote,mFile) values (?,?,?,?,?)"
            self.cursor.execute(sql, (mNo, mMark, mPrice, mNote, mFile))
        except Exception as err:
            print(err)

    def showDB(self):
        #显示爬取数据
        try:
            print("%8s %16s %8s %16s %s" % ("No", "Mark", "Price", "Image", "Note"))
            self.cursor.execute("select mNo,mMark,mPrice,mFile,mNote from phones order by mNo")
            rows = self.cursor.fetchall()
            for row in rows:
                print("%8s %16s %16s %16s %s" % (row[0], row[1], row[2], row[3], row[4]))
        except Exception as err:
            print(err)

    def download (self, src, mFile):
        #下载图像
        try:
            req = urllib.request.Request(src, headers=MySpider.
```

```python
headers)
                    data = urllib.request.urlopen(req, timeout=400)
                    data = data.read()
                    fobj = open(MySpider.imagePath + "\\" + mFile, "wb")
                    fobj.write(data)
                    fobj.close()
            except Exception as err:
                print(mFile + " " + str(err) + " src=" + src)

    def processSpider(self):
        #爬取过程函数
        try:
            print(self.driver.current_url)
            divs = self.driver.find_elements_by_xpath("//div[@class='phone']")
            for div in divs:
                #爬取数据
                src = div.find_element_by_xpath(".//div[position()=2]//img").get_attribute("src")
                src = urllib.request.urljoin(self.driver.current_url, src)
                mark = div.find_element_by_xpath(".//div[position()=3]").text
                price = div.find_element_by_xpath(".//div[position()=4]").text
                note = div.find_element_by_xpath(".//div[position()=5]").text

                self.No = self.No + 1
                no = str(self.No)
                while len(no) < 6:
                    no = "0" + no
                p=src.rfind(".")
                mFile=no+src[p:]
                self.insertDB(no, mark, price, note, mFile)

                #启动下载图像子线程
                T=threading.Thread(target=self.download,args=[src, mFile])
                T.start()
                self.threads.append(T)

            #实现翻页
            try:
                link = self.driver.find_elements_by_xpath("//div[@class='paging']//input[@type='button']")[-2]
                if link.is_enabled():
                    link.click()
                    time.sleep(0.1)
                    self.processSpider()
            except Exception as err:
                print(err)
```

```
            except Exception as err:
                print(err)

    def executeSpider(self, url):
        #爬取函数
        print("Spider starting......")
        self.startUp()
        self.driver.get(url)
        self.processSpider()
        #等待所有线程结束
        for T in self.threads:
            T.join()
        print("Spider completed......")
        self.showDB()
        self.closeUp()

#主程序
spider = MySpider()
spider.executeSpider("http://127.0.0.1:5000")
```

程序说明如下。

（1）记录的编号。

使用 self.No 记录下载的手机数量，并使用它产生一个长度为 6 位的字符串 no 作为数据库中的关键字，同时下载的图像文件名称 mFile 由 no 与图像的扩展名组成。

（2）爬取手机图像。

在确定了手机的图像地址 src 后，设计一个下载函数 download(self,src,mFile)下载该图像。其中，src 是图像地址，mFile 是图像名称。为了使爬虫程序看起来更像是一个浏览器，下载时使用了浏览器表头 MySpider.headers：

```
headers = {
    "User-Agent": "Mozilla/5.0 (Windows; U; Windows NT 6.0 x64; en-US; rv:1.9pre) Gecko/2008072421 Minefield/3.0.2pre"}
```

同时使用一个线程调用该下载函数，实现异步下载：

```
T=threading.Thread(target=self.download,args=[src,mFile])
T.start()
self.threads.append(T)
```

使用 self.threads 列表记录所有的线程，并在程序结束时等待所有的线程结束，保证图像下载完成。

### 5.7.5 执行爬虫程序

先执行服务器程序，再执行爬虫程序，可以看到成功爬取了所有手机的数据与图像。下面是程序执行翻页的过程：

```
Spider starting......
http://127.0.0.1:5000/
http://127.0.0.1:5000/?pageIndex=2
http://127.0.0.1:5000/?pageIndex=3
http://127.0.0.1:5000/?pageIndex=4
http://127.0.0.1:5000/?pageIndex=5
http://127.0.0.1:5000/?pageIndex=6
```

# 项目 ❺ 爬取商城网站数据

```
http://127.0.0.1:5000/?pageIndex=7
http://127.0.0.1:5000/?pageIndex=8
http://127.0.0.1:5000/?pageIndex=9
Spider completed......
```

## 5.8 实战项目 爬取实际商城网站数据

5-8-A 知识讲解　5-8-B 操作演练

 **任务目标**

京东商城网站有大量的商品数据，在搜索文本框中输入某类商品，如"手机"，就可以看到许多手机的信息。本项目使用 selenium 编写一个爬虫程序，在搜索文本框中输入"手机"，自动翻页爬取所有手机的数据与图像，并将数据保存到数据库。

### 5.8.1 解析网站的 HTML 代码

京东商城网站有很多手机信息，用 Chrome 浏览器访问京东商城网站，在搜索文本框中输入"手机"，就会看到图 5-8-1 所示的页面。

将鼠标指针移至手机上，单击鼠标右键，在弹出的快捷菜单中选择"检查"命令，可以看到手机的信息是包含在一组<li>元素中的，再仔细分析，这些<li>都包含在一个<div id="J_goodsList">中，而且每个<li>的格式都是<li class="gl-item">，因此分析<li>中的结构就可以找到手机的各种信息，网站的 HTML 代码如图 5-8-2 所示。

图 5-8-1 在京东商城网站搜索"手机"后的页面

图 5-8-2 网站的 HTML 代码

复制其中一个<li>的代码，如下：

```
    <li class="gl-item" data-pid="100000349372" data-sku="100000503295" data-spu="100000349372">
      <div class="gl-i-wrap">
       <div class="p-img">
        <a href="https://ccc-x.jd.com/dsp/nc?ext=aHR0cHM6Ly9pdGVtLmpkLmNvbS8xMDAwMDA1MDMyOTUuaHRtbA&amp;log=GGwsfgzPEa6GsUjtjtmIPqCcsJJ8_FyzP-oiMm40ZPdMyy4BUaQho5X09iHiOTaBXb5lRyCbqnguIgWB4gciEzLOs2zl26geClF2Ioq8CClJghvlPDDTFItXp_NNt4-Irh4DErYAGKCBfUN09WUKYdtWrjpnRnJzb2V06AwJIRtBYFC-670vScpMxUHRUtgYyT9BrMur50VuMqVl0Wi74c3P67g8CKz0EbDQpSCm87-mn824KGDGtIUHTIfqbVGrclUkU_7haMxMz-jxV2maXdj9qSWTjnXT1Beoc0hl3xOiYxu8131zQ_raOiNa17DKxdTdfLioFnhNlnixJG9MC-tR0BnWMBIBl52Q_kZSuD9e7BLIgid8tNR9ANb4eOUhOXG7eku-06_mEkLhqEuxvdXN2Pk-agFcIOULzpWulu4yP_VZz64ghL_hujkJ8JN4x_zGH_xTNSomO__PGBEse5engWL0paP3EunPiVLDjNZQgXfKwcvQ6cbuYsveUCzuizkO47q-Ra5-ZuK2iu2R5w&amp;v=404" onclick="searchlog(1,100000503295,0,2,'','adwClk=1')" target="_blank" title="【品质特惠】限时优惠200！成交价1499！潮流镜面渐变色，2400万自拍旗舰，骁龙660AIE处理器！小米爆品特惠，选品质，购小米！">
         <img class="" data-img="1" data-lazy-img="done" height="220" source-data-lazy-img="" src="//img10.360buyimg.com/n7/jfs/t1/2617/6/6143/237736/5ba1f42aE71124526/e242e3e39ec95d66.jpg" width="220"/>
        </a>
        <div data-catid="655" data-done="1" data-lease="" data-presale="" data-venid="1000004123">
        </div>
       </div>
       <div class="p-scroll">
        <span class="ps-prev">
         &lt;
        </span>
        <span class="ps-next">
         &gt;
        </span>
        <div class="ps-wrap">
         <ul class="ps-main">
          <li class="ps-item">
           <a class="curr" href="javascript:;" title="梦幻蓝">
            <img class="" data-done="1" data-img="1" data-lazy-img="done" data-presale="" data-sku="100000503295" data-url="https://ccc-x.jd.com/dsp/nc?ext=aHR0cHM6Ly9pdGVtLmpkLmNvbS8xMDAwMDA1MDMyOTUuaHRtbA&amp;log=GGwsfgzPEa6GsUjtjtmIPqCcsJJ8_FyzP-oiMm40ZPdMyy4BUaQho5X09iHiOTaBXb5lRyCbqnguIgWB4gciEzLOs2zl26geClF2Ioq8CClJghvlPDDTFItXp_NNt4-Irh4DErYAGKCBfUN09WUKYdtWrjpnRnJzb2V06AwJIRtBYFC-670vScpMxUHRUtgYyT9BrMur50VuMqVl0Wi74c3P67g8CKz0EbDQpSCm87-mn824KGDGtIUHTIfqbVGrclUkU_7haMxMz-jxV2maXdj9qSWTjnXT1Beoc0hl3xOiYxu8131zQ_raOiNa17DKxdTdfLioFnhNlnixJG9MC-tR0BnWMBIBl52Q_kZSuD9e7BLIgid8tNR9ANb4eOUhOXG7eku-06_mEkLhqEuxvdXN2Pk-agFcIOULzpWulu4yP_VZz64ghL_hujkJ8JN4x_zGH_xTNSomO__PGBEse5engWL0paP3EunPiVLDjNZQgXfKwcvQ6cbuYsveUCzuizkO47q-Ra5-ZuK2iu2R5w&amp;v=404" height="25" src="//img10.360buyimg.com/n7/jfs/t1/2617/6/6143/237736/5ba1f42aE71124526/e242e3e39ec95d66.jpg" width="25"/>
           </a>
          </li>
         </ul>
        </div>
       </div>
       <div class="p-price">
```

```
            <strong class="J_100000503295" data-done="1">
              <em>
                ¥
              </em>
              <i>
                1499.00
              </i>
            </strong>
          </div>
          <div class="p-name p-name-type-2">
            <a href="https://ccc-x.jd.com/dsp/nc?ext=
aHR0cHM6Ly9pdGVtLmpkLmNvbS8xMDAwMDA1MDMyOTUuaHRtbA&amp;log=GGwsfgzPEa6GsUjt
jtmIPqCcsJJ8_FyzP-oiMm40ZPdMyy4BUaQho5X09iHiOTaBXb5lRyCbqnguIgWB4gciEzLOs2z
l26geClF2Ioq8CClJghvlPDDTFItXp_NNt4-Irh4DErYAGKCBfUN09WUKYdtWrjpnRnJzb2V06A
wJIRtBYFC-670vScpMxUHRUtgYyT9BrMur50VuMqVl0Wi74c3P67g8CKz0EbDQpSCm87-mn824K
GDGtIUHTIfqbVGrclUkU_7haMxMz-jxV2maXdj9qSWTjnXT1Beoc0hl3xOiYxu8131zQ_raOiNa
17DKxdTdfLioFnhNlnixJG9MC-tR0BnWMBIBl52Q_kZSuD9e7BLIgid8tNR9ANb4eOUhOXG7eku
-06_mEkLhqEuxvdXN2Pk-agFcIOULzpWulu4yP_VZz64ghL_hujkJ8JN4x_zGH_xTNSomO__PGB
Ese5engWL0paP3EunPiVLDjNZQgXfKwcvQ6cbuYsveUCzuizkO47q-Ra5-ZuK2iu2R5w&amp;v=
404" onclick="searchlog(1,100000503295,0,1,'','adwClk=1')" target="_blank"
title="【品质特惠】限时优惠 200！成交价 1499！潮流镜面渐变色，2400 万自拍旗舰，骁龙 660AIE
处理器！小米爆品特惠，选品质，购小米！">
              <em>
                小米 8 青春版 镜面渐变 AI 双摄 6GB+64GB 梦幻蓝 骁龙 全网通 4G 双卡双待 全面屏拍
照游戏智能
                <font class="skcolor_ljg">
                  手机
                </font>
              </em>
              <i class="promo-words" id="J_AD_100000503295">
                【品质特惠】限时优惠 200！成交价 1499！潮流镜面渐变色，2400 万自拍旗舰，骁龙 660AIE
处理器！小米爆品特惠，选品质，购小米！
              </i>
            </a>
          </div>
          <div class="p-commit" data-done="1">
            <strong>
              <a href="https://ccc-x.jd.com/dsp/nc?ext=
aHR0cHM6Ly9pdGVtLmpkLmNvbS8xMDAwMDA1MDMyOTUuaHRtbA&amp;log=GGwsfgzPEa6GsUjt
jtmIPqCcsJJ8_FyzP-oiMm40ZPdMyy4BUaQho5X09iHiOTaBXb5lRyCbqnguIgWB4gciEzLOs2z
l26geClF2Ioq8CClJghvlPDDTFItXp_NNt4-Irh4DErYAGKCBfUN09WUKYdtWrjpnRnJzb2V06A
wJIRtBYFC-670vScpMxUHRUtgYyT9BrMur50VuMqVl0Wi74c3P67g8CKz0EbDQpSCm87-mn824K
GDGtIUHTIfqbVGrclUkU_7haMxMz-jxV2maXdj9qSWTjnXT1Beoc0hl3xOiYxu8131zQ_raOiNa
17DKxdTdfLioFnhNlnixJG9MC-tR0BnWMBIBl52Q_kZSuD9e7BLIgid8tNR9ANb4eOUhOXG7eku
-06_mEkLhqEuxvdXN2Pk-agFcIOULzpWulu4yP_VZz64ghL_hujkJ8JN4x_zGH_xTNSomO__PGB
Ese5engWL0paP3EunPiVLDjNZQgXfKwcvQ6cbuYsveUCzuizkO47q-Ra5-ZuK2iu2R5w&amp;v=
404" id="J_comment_100000503295" onclick="searchlog(1,100000503295,0,3,'',
'adwClk=1')" target="_blank">
                19 万+
              </a>
              条评价
            </strong>
```

```html
        </div>
        <div class="p-focus">
         <a class="J_focus" data-sku="100000503295" href="javascript:;" onclick="searchlog(1,100000503295,0,5,'','adwClk=1')" title="单击关注">
          <i>
          </i>
          关注
         </a>
        </div>
        <div class="p-shop" data-done="1" data-reputation="99" data-score="0" data-selfware="1" data-verderid="1000004123">
         <span class="J_im_icon">
          <a href="//mall.jd.com/index-1000004123.html" onclick="searchlog(1,1000004123,0,58)" target="_blank" title="小米京东自营旗舰店">
           小米京东自营旗舰店
          </a>
          <b class="im-02" onclick="searchlog(1,1000004123,0,61)" style="background:url(//img14.360buyimg.com/uba/jfs/t26764/156/1205787445/713/9f715eaa/5bc4255bN0776eea6.png) no-repeat;" title="联系客服">
          </b>
         </span>
        </div>
        <div class="p-icons" data-done="1" id="J_pro_100000503295">
         <i class="goods-icons J-picon-tips J-picon-fix" data-idx="1" data-tips="京东自营，品质保障">
          自营
         </i>
        </div>
        <span class="p-promo-flag">
         广告
        </span>
        <img src="https://im-x.jd.com/dsp/np?log=GGwsfgzPEa6GsUjtjtmIPqCcsJJ8_FyzP-oiMm40ZPdMyy4BUaQho5X09iHiOTaBXb5lRyCbqnguIgWB4gciEzLOs2zl26geClF2Ioq8CClJghvlPDDTFItXp_NNt4-Irh4DErYAGKCBfUN09WUKYdtWrjpnRnJzb2V06AwJIRtBYFC-670vScpMxUHRUtgYyT9BrMur50VuMqVl0Wi74bCEGdD_q75HGSzPQGw6hJ4HlaZ8W28vNYOfkLpxilFoRc4qcGHvvWHOiOzImfnEepOqdi4fDVK_-FB3_3oJQA7sQmNUBqdsLNuuYfNe9a4MlImbykNTzzAKxOD42W8vJO1zfl2NpV7UMZA9eUBGv4rKFVn4PYB-WCFzT6SzeH1LUR3ZjJuNp38LQ2ZGFpK_JHhuYDLPgucuEtpSdIUJSV7NaX-VyBg6-eg5oqoIYlFaxfqxYYqR5dGOYIBY_W2CuGdoNvynF09oa4nBudXWOdMUJ4GG-KQ0U6vmwU4IuCEsywY-rl2-sAgHoxJLklZwvpk2XpJ8VXYcVbHTPFEZzU0~&amp;v=404&amp;rt=3" style="display:none;"/>
        <img class="err-poster" source-data-lazy-advertisement="done" src="//misc.360buyimg.com/lib/img/e/blank.gif" style="display: none;"/>
    </div>
  </li>
```

## 5.8.2 爬取网站数据

### 1. 爬取手机数据

从 HTML 代码中可以看到，每个<li>都包含在<div id="J_goodsList">中，而且每个<li>的格式都为<li class="gl-item">。因此使用：

```
    lis = self.driver.find_elements_by_xpath("//div[@id='J_goodsList']
//li[@class='gl-item']")
    for li in lis:
        # 从<li>爬取数据
```
通过循环得到每个<li>元素,每部手机的数据从<li>元素中进一步爬取。

### 2. 爬取价格

价格信息包含在<div class='p-price'>下面的<i>元素中,因此价格 price 可以这样爬取:

```
try:
    price = li.find_element_by_xpath(".//div[@class='p-price']//i").text
except:
    price = "0"
```

### 3. 爬取品牌与简介

品牌与简介信息包含在<div class='p-name p-name-type-2'>下面的<em>元素中,而且品牌是<em>下面文本中的第一部分(用空格分开),因此品牌 mark 与简介 note 可以这样爬取:

```
try:
    note = li.find_element_by_xpath(".//div[@class='p-name p-name-type-
2']//em").text
    mark = note.split(" ")[0]
    mark = mark.replace("爱心东东\n", "")
    mark = mark.replace(",", "")
    note = note.replace("爱心东东\n", "")
    note = note.replace(",", "")
except:
    note = ""
    mark = ""
```

### 4. 爬取图像地址

仔细分析 HTML 代码,可以看到手机图像包含在<div class="p-img">下面的<a>的<img>元素中,图像要么存储于<img>的 src 属性中,要么存储于<img>的 data-lazy-img 属性中,因此在这两个属性中取两个地址 src1 与 src2,程序如下:

```
try:
    src1 = li.find_element_by_xpath(".//div[@class='p-img']//a//img").
get_attribute("src")
except:
    src1=""
try:
    src2 = li.find_element_by_xpath(".//div[@class='p-img']//a//img").
get_attribute("data-lazy-img")
except:
    src2=""

#地址 src1 与 src2 中一般有一个图像存在,编写 download()下载函数
def download(self, src1, src2, mFile):
    data = None
    if src1:
        try:
```

```
                req = urllib.request.Request(src1, headers=MySpider.headers)
                resp = urllib.request.urlopen(req, timeout=400)
                data = resp.read()
            except:
                pass
        if not data and src2:
            try:
                req = urllib.request.Request(src2, headers=MySpider.headers)
                resp = urllib.request.urlopen(req, timeout=400)
                data = resp.read()
            except:
                pass
        if data:
            fobj = open(MySpider.imagePath + "\\" + mFile, "wb")
            fobj.write(data)
            fobj.close()
            print("download ", mFile)
```

如果src1存在，download()试图先从src1下载；如果src1不存在或者从src1下载失败，就从src2下载。一般情况下总有一个地址可以下载成功。

### 5.8.3 实现网页翻页

网站中的手机很多，有很多个网页，找到网页的翻页代码，发现翻页不是通过简单的HTML代码控制的，而是通过JavaScript代码控制的，如图5-8-3所示。

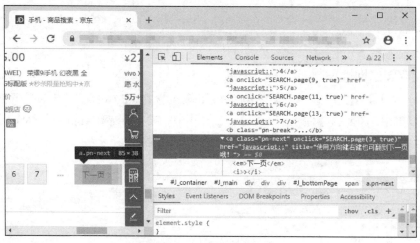

图5-8-3 网页翻页

由此可见，要爬取下一个页面手机的数据和图像，就必须获取控制翻页的超链接元素<a>，并模仿鼠标单击去单击该链接。

复制翻页的HTML代码，如下：

```
<span class="p-num">
 <a class="pn-prev disabled">
  <i>
   &lt;
```

```html
    </i>
    <em>
     上一页
    </em>
   </a>
   <a class="curr" href="javascript:;">
    1
   </a>
   <a href="javascript:;" onclick="SEARCH.page(3, true)">
    2
   </a>
   <a href="javascript:;" onclick="SEARCH.page(5, true)">
    3
   </a>
   <a href="javascript:;" onclick="SEARCH.page(7, true)">
    4
   </a>
   <a href="javascript:;" onclick="SEARCH.page(9, true)">
    5
   </a>
   <a href="javascript:;" onclick="SEARCH.page(11, true)">
    6
   </a>
   <a href="javascript:;" onclick="SEARCH.page(13, true)">
    7
   </a>
   <b class="pn-break">
    ……
   </b>
   <a class="pn-next" href="javascript:;" onclick="SEARCH.page(3, true)" title="使用方向键右键也可翻到下一页哦！ ">
    <em>
     下一页
    </em>
    <i>
     &gt;
    </i>
   </a>
  </span>
```

由此可见，只要找到<span class="p-num">，然后找到"下一页"的超链接，就能实现翻页。在正常翻页时超链接是<a class='pn-next'>，到最后一页不能翻页时超链接变成<a class='pn-next disabled'>。因此编写下列程序找到nextPage就能实现翻页：

```
try:
    self.driver.find_element_by_xpath("//span[@class='p-num']
//a[@class='pn-next disabled']")
except:
    nextPage = self.driver.find_element_by_xpath("//span[@class='p-num']
//a[@class='pn-next']")
    nextPage.click()
```

如果没有找到<a class='pn-next disabled'>元素就找<a class='pn-next'>元素，然后使用nextPage.click()实现翻页。

## 5.8.4 编写爬虫程序

根据前面的分析,编写爬虫程序 spider.py,如下:

```python
from selenium import webdriver
from selenium.webdriver.chrome.options import Options
import urllib.request
import threading
import sqlite3
import os
import datetime
from selenium.webdriver.common.keys import Keys
import time

class MySpider:
    headers = {
        "User-Agent": "Mozilla/5.0 (Windows; U; Windows NT 6.0 x64; en-US; rv:1.9pre) Gecko/2008072421 Minefield/3.0.2pre"}
    imagePath = "download"

    def startUp(self,url,key):
        # Initializing Chrome browser
        chrome_options = Options()
        chrome_options.add_argument('--headless')
        chrome_options.add_argument('--disable-gpu')
        self.driver = webdriver.Chrome(chrome_options=chrome_options)
        self.driver.maximize_window()

        # Initializing variables
        self.threads = []
        self.No = 0
        self.imgNo=0

        # Initializing database
        try:
            self.con = sqlite3.connect("phones.db")
            self.cursor = self.con.cursor()
            try:
                # 如果有表就删除
                self.cursor.execute("drop table phones")
            except:
                pass
            try:
                # 建立新的表
                sql = "create table phones (mNo varchar(32) primary key,mMark varchar(256),mPrice varchar(32),mNote varchar(1024),mFile varchar(256))"
                self.cursor.execute(sql)
            except:
                pass
        except Exception as err:
            print(err)
```

```python
            # Initializing images folder
            try:
                if not os.path.exists(MySpider.imagePath):
                    os.mkdir(MySpider.imagePath)
                images = os.listdir(MySpider.imagePath)
                for img in images:
                    s = os.path.join(MySpider.imagePath, img)
                    os.remove(s)
            except Exception as err:
                print(err)

            #网页第一页,输入key后跳转到新的页面
            self.driver.get(url)
            keyInput=self.driver.find_element_by_id("key")
            keyInput.send_keys(key)
            keyInput.send_keys(Keys.ENTER)

    def closeUp(self):
        try:
            self.con.commit()
            self.con.close()
            self.driver.close()
        except Exception as err:
            print(err);

    def insertDB(self, mNo, mMark, mPrice, mNote, mFile):
        try:
            sql = "insert into phones (mNo,mMark,mPrice,mNote,mFile) values (?,?,?,?,?)"
            self.cursor.execute(sql, (mNo, mMark, mPrice, mNote, mFile))
        except Exception as err:
            print(err)

    def showDB(self):
        try:
            con=sqlite3.connect("phones.db")
            cursor=con.cursor()
            print("%-8s %-16s %-8s %-16s %s" % ("No", "Mark", "Price", "Image", "Note"))
            cursor.execute("select mNo,mMark,mPrice,mFile,mNote from phones order by mNo")
            rows = cursor.fetchall()
            for row in rows:
                print("%-8s %-16s %-8s %-16s %s" % (row[0], row[1], row[2], row[3], row[4]))
            con.close()
        except Exception as err:
            print(err)

    def download(self, src1,src2,mFile):
```

```python
            #下载图像，先从src1地址下载，下载失败时再从src2地址下载
            data=None
            if src1:
                try:
                    req = urllib.request.Request(src1, headers=MySpider.headers)
                    resp = urllib.request.urlopen(req, timeout=400)
                    data = resp.read()
                except:
                    pass
            if not data and src2:
                try:
                    req = urllib.request.Request(src2, headers=MySpider.headers)
                    resp = urllib.request.urlopen(req, timeout=400)
                    data = resp.read()
                except:
                    pass
            if data:
                fobj = open(MySpider.imagePath + "\\" + mFile, "wb")
                fobj.write(data)
                fobj.close()
                print("download ",mFile)

    def processSpider(self):
        #爬取一个页面的数据
        try:
            time.sleep(1)
            #等待1s
            print(self.driver.current_url)
            lis = self.driver.find_elements_by_xpath("//div[@id='J_goodsList']//li[@class='gl-item']")
            for li in lis:
                # We find that the image is either in src or in data-lazy-img attribute
                try:
                    src1 = li.find_element_by_xpath(".//div[@class='p-img']//a//img").get_attribute("src")
                except:
                    src1=""
                try:
                    src2 = li.find_element_by_xpath(".//div[@class='p-img']//a//img").get_attribute("data-lazy-img")
                except:
                    src2=""
                try:
                    price = li.find_element_by_xpath(".//div[@class='p-price']//i").text
                except:
                    price="0"
                try:
```

```python
                        note = li.find_element_by_xpath(".//div
[@class='p-name p-name-type-2']//em").text
                        mark = note.split(" ")[0]
                        mark = mark.replace("爱心东东\n", "")
                        mark = mark.replace(",", "")
                        note = note.replace("爱心东东\n", "")
                        note = note.replace(",", "")
                    except:
                        note=""
                        mark=""

                    self.No = self.No + 1
                    no = str(self.No)
                    while len(no) < 6:
                        no = "0" + no
                    print(no,mark,price)
                    #爬取图像地址
                    if src1:
                        src1=urllib.request.urljoin(self.driver.current_url,src1)
                        p = src1.rfind(".")
                        mFile = no + src1[p:]
                    elif src2:
                        src2=urllib.request.urljoin(self.driver.current_url,src2)
                        p = src2.rfind(".")
                        mFile = no + src2[p:]
                    if src1 or src2:
                        #启动子线程，下载图像
                        T = threading.Thread(target=self.download,args=(src1,src2,mFile))
                        T.setDaemon(False)
                        T.start()
                        self.threads.append(T)
                    else:
                        mFile = ""
                    self.insertDB(no, mark, price, note, mFile)

                #实现翻页
                try:
                    #如果这个无效按钮存在，就跳转到最后一页
                    self.driver.find_element_by_xpath("//span[@class='p-num']//a[@class='pn-next disabled']")
                except:
                    #找到"下一页"按钮，单击该按钮跳转到下一页
                    nextPage = self.driver.find_element_by_xpath("//span[@class='p-num']//a[@class='pn-next']")
                    nextPage.click()
                    self.processSpider()
            except Exception as err:
```

```
                    print(err)

        def executeSpider(self, url, key):
            #爬取函数
            starttime = datetime.datetime.now()
            print("Spider starting......")
            self.startUp(url,key)
            self.processSpider()
            self.closeUp()
            #等待线程结束
            for t in self.threads:
                t.join()
            print("Spider completed......")
            endtime = datetime.datetime.now()
            elapsed = (endtime - starttime).seconds
            print("Total ", elapsed, " seconds elapsed")

#主程序
url = "http://www.jd.com"
spider = MySpider()
while True:
    print("1.爬取")
    print("2.显示")
    print("3.退出")
    s=input("请选择(1,2,3):")
    if s=="1":
        spider.executeSpider(url,"手机")
    elif s=="2":
        spider.showDB()
    elif s=="3":
        break
```

startUp()函数负责初始化数据库 phones.db 并建立一张空的 phones 表以存储数据，同时创建 download 文件夹并清空文件夹中的文件以存储下载的图像。该函数还负责查找网页的 <input id="key">，模拟键盘输入要爬取的商品关键字 key，并模拟按 Enter 键后跳转到商品的网页，关键程序如下：

```
keyInput=self.driver.find_element_by_id("key")
keyInput.send_keys(key)
keyInput.send_keys(Keys.ENTER)
```

### 5.8.5　执行爬虫程序

执行爬虫程序，耗时 921s，总共爬取到近 100 个页面的近 6000 部手机的数据与图像，下面是部分结果。

```
Spider starting......
https://search.jd.com/Search?keyword=%E6%89%8B%E6%9C%BA&enc=utf-8&pvid=13986d0072404844b2dae1417d9191b0
000001 Apple 6299.00
000002 OPPO 1699.00
```

```
000003 一加手机 6T  3599.00
000004 荣耀 10 青春版 1399.00
000005 荣耀 10  2199.00
000006 小米 8  2299.00
000007 荣耀 8X  1399.00
000008 Apple  3998.00
...
download  005993.jpg
005994 华为（HUAWEI） 1799.00
download  005994.jpg
005995 华为（HUAWEI） 828.00
download  005995.jpg
005996 华为  3388.00
No       Mark              Price     Image          Note
download  005996.jpg
Spider completed......
Total  921  seconds elapsed
```

图 5-8-4 所示是 phones.db 数据库中的程序爬取到的数据，图 5-8-5 所示是爬取并存储到 download 文件夹中的近 6000 幅图像。

图 5-8-4　爬取到的数据

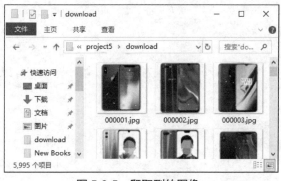

图 5-8-5　爬取到的图像

# Python 爬虫项目教程（微课版）

## 项目总结

这个项目涉及一个有多个网页的商城网站，我们使用 selenium 爬取各个网页的数据，实现了爬取商城网站数据的爬虫程序。

selenium 是一个优秀的程序框架，它可以搭配多个浏览器的驱动程序，完全模拟实际的浏览器，支持静态网页与 JavaScript 程序控制的动态网页的数据爬取，能在程序中有效地执行 JavaScript 程序。它还支持 XPath、CSS 等多种查找元素与数据的方法，能很好地操控各个网页元素，如模拟键盘输入与鼠标单击等。因此使用 selenium 能编写出功能强大的爬虫程序。

## 练习 5

1. 说明 selenium 的工作原理。为什么其能执行网页中的 JavaScript 程序？
2. 说明 selenium 的等待方式。各种等待方式有什么特点？
3. 启动如下的服务器程序。

```
import flask
app=flask.Flask(__name__)
@app.route("/")
def index():
    s='''<body>
<span id="jMsg"></span>
</body>
<script>
    document.getElementById("jMsg").innerHTML="javascript 信息";
</script>
'''
    return s
app.run()
```

（1）用 urllib 编写程序来获取网站的 HTML 代码。
（2）用 selenium 编写程序来获取网站的 HTML 代码。
通过这两种方法获取的 HTML 代码是否相同？为什么？

4. 启动如下的服务器程序。

```
import flask
app=flask.Flask(__name__)
@app.route("/")
def index():
    s='''<body>
<span id="jMsg"></span>
</body>
<script>
    function msg()
    {
      document.getElementById("jMsg").innerHTML="javascript 信息";
    }
    window.setTimeout(msg,1000);
</script>
'''
```

```
        return s
    app.run()
```
试用 selenium 编写爬虫程序等待<span id="jMsg">…</span>中出现的字符串，并爬取该字符串。

5. 启动如下的服务器程序。

```
import flask
app=flask.Flask(__name__)
@app.route("/")
def index():
    s='''<body>
<form name='frm' action='/handin' method='get'>
姓名<input type='text' name='name'><br>
性别<input type='radio' name='gender' value='男'>男<input type='radio' name='gender' value='女'>女<br>
<input type='submit' value='提交' >
</form>
'''
    return s

@app.route("/handin")
def handin():
    name=flask.request.values.get("name")
    gender = flask.request.values.get("gender")
    return "<div id='name'>"+name+"</div><div id='gender'>"+gender+"</div>"

app.run()
```

试用 selenium 编写程序：
（1）填写姓名和性别为"张三"与"女"，并提交；
（2）爬取提交后的姓名与性别。

# 项目 6 爬取景区网站数据

拓展阅读

描写祖国大好河山的古诗

爬虫程序爬取的数据有时候是不规则的,这种不规则的数据往往很难使用传统的关系数据库(如 SQLite、MySQL 等)进行存储。要存储这样的数据,一个比较好的方法是使用 NoSQL 数据库。亚马逊云服务(Amazon Web Service,AWS)的 DynamoDB 数据库就是一个不错的选择。在本项目中,我们将学习如何使用 AWS 的 DynamoDB 数据库存储这种不规则的数据。

本项目将爬取我国主要的旅游景区的数据,我国地大物博,人杰地灵,有许多的名胜古迹、人文景观。有广阔无垠的草原,巍巍耸立的高山,气势磅礴的黄河长江,绵延万里的长城等。通过本项目,大家可以领略祖国的壮丽河山,感悟中华大地的文化遗产,增强我们的民族自豪感。

6-1-A 6-1-B

知识讲解　操作演练

## 6.1 项目任务

中国旅游信息网是一个含有丰富的旅游资源数据的网站,进入该网站可以查看各种各样的旅游信息,例如,选择查看 5A 景区,就可以看到全国 5A 景区的介绍,如图 6-1-1 所示。

在爬取这个网站的数据之前,先学习爬取模拟景区网站的数据。创建的模拟景区网站如图 6-1-2 所示,它展现了两个景区的信息,包括景区名称、景区类型、景区资质、景区级别、酒店、交通等数据。我们通过这个模拟景区网站来学习数据爬取与云存储的技术。

图 6-1-1　景区信息

图 6-1-2　模拟景区网站

# 项目 ❻ 爬取景区网站数据

## 6.2 使用DynamoDB存储模拟景区网站数据

6-2-A
知识讲解

6-2-B
操作演练

### 任务目标

首先创建一个有两个景区的景区网站，编写爬虫程序爬取网站的数据，并把数据存储在数据库中。在本节中，我们引入AWS的DynamoDB数据库，学习如何存储不规则的数据。

### 6.2.1 创建模拟景区网站

#### 1. 创建网站模板

根据中国旅游信息网站模板文件，创建如下 scene.html 模板文件：

```
<style>
    .sightimg {display:inline-block; width:200px; vertical-align:top}
    .sightdetail {display:inline-block;}
    .sightbase {margin:10px;display:inline-block;}
    .sighthotel {margin:10px;display:inline-block;}
    li {list-style: none; margin:8px;}
</style>

<body>
<ul>
    <li>
        <div>
    <div class="sightimg"><img alt="双月湾风景区" height="180" src="static/双月湾.jpg" width="180"/></div>
    <div class="sightdetail">
     <h3>双月湾风景区</h3>
     <ul class="sightbase">
      <li>景区类型：<span>海滨海岛</span><span>生物景观</span></li>
      <li>景区资质：<span>国家级风景名胜区</span></li>
      <li>景区级别：<span>4A</span></li>
       <li>适合季节：<span>四季皆宜</span></li>
     </ul>
     <div class="sighthotel">
     <ul ><li>金海湾喜来登度假酒店:<span>¥700</span></li>
       <li>双月湾虹海湾假日酒店:<span>¥400</span></li>
     </ul>
    </div>
    <h6>景区全攻略:
        <div>[辖内景区]:<span>双月湾沙滩浴场</span> <span>海龟基地</span> <span>山顶观景台</span></div>
            <div>[交通概况]:<span>深汕高速-稔山出口-左转-大浦垅-右转-双月湾</span><br>
            <span>惠东-双月湾：约20分钟一班</span>
        </div>
```

```html
                </h6>
            </div>
                </div>
        </li>

        <li>
            <div>
                <div class="sightimg"><img alt="深圳华侨城旅游度假区" height="180" src="static/东部华侨城.jpg" width="180"/></div>
                <div class="sightdetail">
                    <h3>深圳华侨城旅游度假区</h3>
                    <ul class="sightbase">
                        <li>景区类型：<span>城市风景</span><span>生态景观</span></li>
                        <li>景区资质：<span>国家级风景名胜区</span></li>
                        <li>景区级别：<span>5A</span></li>
                        <li>适合季节：<span>四季皆宜</span></li>
                    </ul>
                    <div class="sighthotel">
                    <ul ><li>雅枫宾馆（深圳华侨城店）:<span>￥1700</span></li>
                        <li>深圳华侨城洲际大酒店:<span>￥1400</span></li>
                            <li>深圳华侨城蓝汐精品酒店:<span>￥2400</span></li>
                    </ul>
                    </div>
                    <h6>景区全攻略：
                        <div>[辖内景区]:<span>欢乐海岸</span> <span>锦绣中华</span></div>
                            <div>[交通概况]:<span>地处华侨城风景区，离华侨城地铁口仅50米；离市中心15千米，10分钟车程</span></div>
                    </h6>
            </div>
                </div>
        </li>
    </ul>
    </body>
```

#### 2．创建网站服务器程序

在项目中创建服务器程序 server.py，它的作用是返回 templates 文件夹中的 scene.html 文件，程序如下：

```python
import flask
app=flask.Flask(__name__)
@app.route("/")
def index():
    return flask.render_template("scene.html")
app.run()
```

运行服务器程序后，使用浏览器访问"http://127.0.0.1:5000"，就会看到图 6-1-2 所示的网页。

## 项目 ❻  爬取景区网站数据

### 6.2.2 爬取网站数据

首先获取 HTML 代码，建立 BeautifulSoup 对象 soup：

```
resp=urllib.request.urlopen(url)
html=resp.read().decode()
soup=BeautifulSoup(html,"lxml")
```

#### 1. 爬取景区名称

景区信息包含在<ul>的直接子节点<li>中，因此可以先获取所有的<ul>，再循环遍历所有的<li>，建立的 div 对象是<div class="sightdetail">元素，景区名称包含在它下面的<h3>中，因此景区名称 sName 可以这样获取：

```
lis=soup.find("ul").find_all("li",recursive=False)
for li in lis:
    div=li.find("div",attrs={"class":"sightdetail"})
    sName=div.find("h3").text
```

#### 2. 爬取景区类型、景区资质、景区级别、适合季节

这些数据包含在<ul>下面的 4 个<li>中，先获取这 4 个<li>，然后分别获取这些数据：

```
list=div.find("ul").find_all("li")
sType=[]
for sp in list[0].find_all("span"):
    sType.append(sp.text)
sSource=list[1].find("span").text
sLevel=list[2].find("span").text
sTime=list[3].find("span").text
```

其中，sType 是景区类型，它包含在第一个<li>的多个<span>中，因此把它设计成一个列表。

#### 3. 爬取景区酒店

景区酒店信息包含在<div class="sighthotel">的<li>中，每个<li>都有两个部分，例如：

```
<li>双月湾虹海湾假日酒店:<span>¥400</span></li>
```

第一部分的酒店名称是一个文本，可以设计一个函数 getFirstText()来获取：

```
def getFirstText(self,li):
    for c in li.children:
        if isinstance(c,bs4.element.NavigableString):
            d=c.string.replace(" ","").replace("\n","")
            if d!="":
                return d
    return ""
```

这个函数找到的<li>中第一个非空的文本字符串就是酒店名称，而酒店的价格通过 li.find("span").text 获取。

#### 4. 爬取辖内景区与交通概况

辖内景区信息包含在<h6>的第一个<div>的各个<span>中，交通概况信息包含在第二个<div>的各个<span>中，它们都是多值的数据，把它们放在一个列表中，因此使用：

```
sScene=[]
for sp in div.find("h6").find_all("div")[0].find_all("span"):
```

```
            sScene.append(sp.text)
        sRoute=[]
        for sp in div.find("h6").find_all("div")[1].find_all("span"):
            sRoute.append(sp.text)
```

就可以获取辖内景区 sScene 与交通概况 sRoute。

### 6.2.3 编写爬虫程序

根据前面的分析,编写爬虫程序 spider.py,如下:

```
import urllib.request
from bs4 import BeautifulSoup
import bs4

def getFirstText(li):
    #获取第一个子节点
    for c in li.children:
        if isinstance(c,bs4.element.NavigableString):
            d=c.string.replace(" ","").replace("\n","")
            if d!="":
                return d
    return ""

def spider(url):
    #爬虫函数
    try:
        resp=urllib.request.urlopen(url)
        html=resp.read().decode()
        soup=BeautifulSoup(html,"lxml")
        lis=soup.find("ul").find_all("li",recursive=False)
        for li in lis:
            div=li.find("div",attrs={"class":"sightdetail"})
            sName=div.find("h3").text
            list=div.find("ul").find_all("li")
            sType=[]
            for sp in list[0].find_all("span"):
                sType.append(sp.text)
            sSource=list[1].find("span").text
            sLevel=list[2].find("span").text
            sTime=list[3].find("span").text
            sHotel=[]
            for h in div.find("div",attrs={"class":"sighthotel"}).find_all("li"):
                ht={}
                ht["name"]=getFirstText(h)
                ht["price"]=h.find("span").text
                sHotel.append(ht)
            sScene=[]
            for sp in div.find("h6").find_all("div")[0].find_all("span"):
                sScene.append(sp.text)
            sRoute=[]
            for sp in div.find("h6").find_all("div")[1].find_all
```

```
("span"):
                        sRoute.append(sp.text)
                print(sName)
                print(sType)
                print(sSource)
                print(sLevel)
                print(sTime)
                print(sHotel)
                print(sScene)
                print(sRoute)
                print()
        except Exception as err:
            print(err)

spider("http://127.0.0.1:5000")
```

## 6.2.4 执行爬虫程序

执行爬虫程序得到如下结果：

双月湾风景区
['海滨海岛', '生物景观']
国家级风景名胜区
4A
四季皆宜
[{'name': '金海湾喜来登度假酒店:', 'price': '¥700'}, {'name': '双月湾虹海湾假日酒店:', 'price': '¥400'}]
['双月湾沙滩浴场', '海龟基地', '山顶观景台']
['深汕高速−稔山出口−左转−大浦坨−右转−双月湾', '惠东−双月湾：约20分钟一班']

深圳华侨城旅游度假区
['城市风景', '生态景观']
国家级风景名胜区
5A
四季皆宜
[{'name': '雅枫宾馆（深圳华侨城店）:', 'price': '¥1700'}, {'name': '深圳华侨城洲际大酒店:', 'price': '¥1400'}, {'name': '深圳华侨城蓝汐精品酒店:', 'price': '¥2400'}]
['欢乐海岸', '锦绣中华']
['地处华侨城风景区，离华侨城地铁口仅50米；离市中心15千米，10分钟车程']

从这个结果可以看到，景区的一些数据是不规则的数据，例如，酒店数据是一个字典，这样的数据比较难直接存储在关系数据库中。实际上，比较好的方法是将这些数据存储在NoSQL数据库中，如AWS的DynamoDB数据库中。

## 6.2.5 DynamoDB简介

AWS是业界很有影响力的云服务。DynamoDB数据库是基于AWS的NoSQL数据库，可以方便地存储任何类型的数据。NoSQL数据库存储数据的基本单位是项目Item，而不是关系数据库中的字段。Item 可以是复杂的数据，基本结构类似字典。例如，把一个景区的数据组织成一个字典：

```
Item={
'sName': '双月湾风景区',
'sLevel': '4A',
'sTime': '四季皆宜',
'sSource': '国家级风景名胜区',
'sType': ['海滨海岛', '生物景观'],
'sHotel': [{'name': '金海湾喜来登度假酒店:', 'price': '¥700'}, {'name': '双月湾虹海湾假日酒店:', 'price': '¥400'}],
'sScene': ['双月湾沙滩浴场', '海龟基地', '山顶观景台'],
'sRoute': ['深汕高速—稔山出口—左转—大浦圩—右转—双月湾', '惠东—双月湾：约 20 分钟一班']
}
```

那么整个 Item 就一次性存储到 DynamoDB 数据库中，读取时可以根据其关键字 sName 一次性读取整个 Item。

## 6.3 登录 AWS 数据库

6-3-A 知识讲解　　6-3-B 操作演练

### 任务目标

要使用 AWS 的 DynamoDB 数据库存储数据，就必须先登录 AWS，然后使用 AWS 的 DynamoDB 数据库创建数据表。在本节中，我们主要学习如何进行这些操作。

### 6.3.1 登录 AWS

AWS 是亚马逊云服务，而 DynamoDB 是 AWS 中提供的一种服务，在使用它们之前首先要有一个 AWS 账号。AWS 推出了一些免费的服务，读者可以免费注册一个 AWS 账号。

使用 AWS 账号进入 AWS，创建一个能操作 DynamoDB 数据库的用户。图 6-3-1 所示为创建的一个名称为"dynamodb.user"的用户，并勾选"编程访问"复选框。这样设置后用户有一个密码，使用密码可以编程访问 AWS。

图 6-3-1　创建 AWS 用户

## 项目 ❻ 爬取景区网站数据

在创建的第一步选择"直接附加现有策略",然后选择"AmazonDynamoDBFullAccess"策略,使得用户具有操作 DynamoDB 数据库的全部权限,如图 6-3-2 所示。创建好的 dynamodb.user 用户如图 6-3-3 所示。

图 6-3-2　选择策略

图 6-3-3　dynamodb.user 用户

这个 dynamodb.user 用户包含一个 Access key ID 和一个 Secret access key,由 AWS 管理机构提供,从用户的"安全证书"下载,如图 6-3-4 所示。下载的结果存储在 accessKeys.csv 文件中,这是一个文本文件,内容如图 6-3-5 所示。

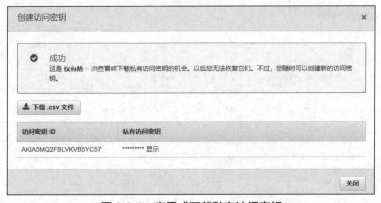

图 6-3-4　查看或下载私有访问密钥

图 6-3-5　accessKeys.csv 文件

这个 accessKeys.csv 文件包含用户 dynamodb.user 的 Access key ID 值与 Secret access key 值，程序访问 DynamoDB 数据库时要使用它们。

### 6.3.2　创建数据库表

要使用 Python 操作 AWS，必须先使用 pip 命令安装 boto3 库，命令是：

```
pip install boto3
```

成功安装 boto3 后，就可以使用 Python 创建 DynamoDB 数据库的表了。

DynamoDB 数据库的表与一般关系数据库的表一样，表中有一个关键字。关系数据库表的关键字是表中一行记录的唯一标识，同样，DynamoDB 数据库表的关键字是表中一个项目的唯一标识。

具体到景区的数据，不妨设置景区名称 sName 为关键字，在 DynamoDB 数据库中以 sName 为关键字创建一个 scenes 表，程序如下：

```python
import boto3
import sys

def readKeys():
    keys={}
    try:
        f=open("accessKeys.csv","rt")
        rows=f.readlines()
        if len(rows)>=2:
            s=rows[1].strip("\n")
            s=s.split(",")
            keys={"keyID":s[0],"secretKey":s[1]}
        f.close()
    except Exception as err:
        print(err)
        sys.exit(0)
    return keys

try:
    keys=readKeys()
    dynamoDB = boto3.resource('dynamodb',
        aws_access_key_id=keys["keyID"],
        aws_secret_access_key=keys["secretKey"],
        region_name="cn-northwest-1")
    table = dynamoDB.create_table(
        TableName="scenes",
        KeySchema=[
            {
                'AttributeName': 'sName',
```

```
                    'KeyType': 'HASH'
                }
            ],
            AttributeDefinitions=[
                {
                    'AttributeName': 'sName',
                    'AttributeType': 'S'
                }
            ],
            ProvisionedThroughput={
                'ReadCapacityUnits': 5,
                'WriteCapacityUnits': 5
            }
        )
        dynamoDB.meta.client.get_waiter('table_exists').wait(TableName=
"scenes")
        print("done!")
    except Exception as err:
        print("errors: ", str(err))
```

程序说明如下。

（1）函数：

readKeys()函数从 accessKeys.csv 文件读取 Access key ID（简称 keyID）与 Secret access key（简称 secretKey）的值。

（2）语句：

```
dynamoDB = boto3.resource('dynamodb',
    aws_access_key_id=keys["keyID"],
    aws_secret_access_key=keys["secretKey"],
    region_name="cn-northwest-1")
```

该语句用于获取 DynamoDB 数据库对象，其中就使用了用户的 keyID 与 secretKey，而 cn-northwest-1 表示数据库的区域，这里指中国西北区。

（3）语句：

```
table = dynamoDB.create_table(...)
```

该语句调用 create_table()函数创建表格，其中要填写一些参数，如下：

① 表的名称通过 TableName="scenes"确定为 scenes。

② 表的关键字通过如下代码设置为 sName，类型是 HASH，即散列值。

```
KeySchema=[
    {
        'AttributeName': 'sName',
        'KeyType': 'HASH'
    }
],
```

③ 表的结构通过如下代码设置为 sName，类型是 S，即字符串。

```
AttributeDefinitions=[
    {
        'AttributeName': 'sName',
        'AttributeType': 'S'
    }
],
```

④ 表的读写特性通过如下代码设置,其值一般使用推荐值。

```
ProvisionedThroughput={
    'ReadCapacityUnits': 5,
    'WriteCapacityUnits': 5
}
```

(4) 语句:

```
dynamoDB.meta.client.get_waiter('table_exists').wait(TableName="scenes")
```

该语句用于等待表创建完成。

程序执行完毕后就在 AWS 的 DynamoDB 数据库中创建了一个名称为 scenes 的表格,我们可以通过 AWS 的后台看到已经创建的 scenes 表,如图 6-3-6 所示。

图 6-3-6　AWS 的 scenes 表

## 6.4　DynamoDB 数据库操作

6-4-A 知识讲解　　6-4-B 操作演练

**任务目标**

数据库的基本操作包含添加、删除、修改、查询等操作。在本节中,我们通过景区数据的存储来学习 DynamoDB 的这些基本操作。

### 6.4.1　存储数据

DynamoDB 使用下面的语句获取 table 表格对象:

```
dynamoDB = boto3.resource('dynamodb',
        aws_access_key_id=keys["keyID"],
        aws_secret_access_key=keys["secretKey"],
        region_name="cn-northwest-1")
table = dynamoDB.Table("scenes")
```

然后 table 表就可以使用 put_item()来存储数据,例如:

```
import boto3
import sys
#省略 readKeys()函数
try:
```

```
        keys=readKeys()
        dynamoDB = boto3.resource('dynamodb',
            aws_access_key_id=keys["keyID"],
            aws_secret_access_key=keys["secretKey"],
            region_name="cn-northwest-1")
        table = dynamoDB.Table("scenes")
        table.put_item(
            Item={
                'sName': '双月湾风景区',
                'sLevel': '4A',
                'sTime': '四季皆宜',
                'sSource': '国家级风景名胜区',
                'sType': ['海滨海岛', '生物景观'],
                'sHotel': [{'name': '金海湾喜来登度假酒店:', 'price': '¥700'},
{'name': '双月湾虹海湾假日酒店:', 'price': '¥400'}],
                'sScene': ['双月湾沙滩浴场', '海龟基地', '山顶观景台'],
                'sRoute': ['深汕高速−稔山出口−左转−大浦垃−右转−双月湾', '惠东−双月
湾: 约20分钟一班']
            }
        )
    print("done!")
except Exception as err:
    print("errors: ", str(err))
```

执行该程序后显示"done!",就表明数据已经存储到数据库中。使用 AWS 后台管理程序可以看到名称为"双月湾风景区"的数据项目,如图 6-4-1 所示。

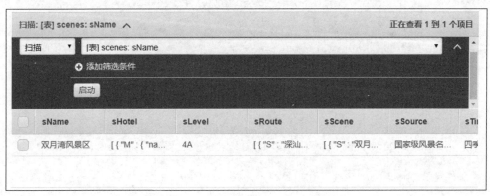

图 6-4-1 存储在 AWS 中的数据

## 6.4.2 读取数据

使用 get_item()读取一个数据项目。使用 get_item()时必须指定要读取的项目的关键字 Key, 例如:

```
import boto3
import sys
#省略 readKeys()函数
try:
    keys=readKeys()
```

```
            dynamoDB = boto3.resource('dynamodb',
                aws_access_key_id=keys["keyID"],
                aws_secret_access_key=keys["secretKey"],
                region_name="cn-northwest-1")
            table = dynamoDB.Table("scenes")
            resp=table.get_item(Key={'sName': '双月湾风景区'})
            print(resp["Item"])
    except Exception as err:
            print("errors: ", str(err))
```

执行该程序，结果如下。

{'sLevel': '4A', 'sName': '双月湾风景区', 'sScene': ['双月湾沙滩浴场', '海龟基地', '山顶观景台'], 'sRoute': ['深汕高速-稔山出口-左转-大浦圩-右转-双月湾', '惠东-双月湾：约20分钟一班'], 'sHotel': [{'name': '金海湾喜来登度假酒店:', 'price': '¥700'}, {'name': '双月湾虹海湾假日酒店:', 'price': '¥400'}], 'sType': ['海滨海岛', '生物景观'], 'sTime': '四季皆宜', 'sSource': '国家级风景名胜区'}

### 6.4.3 修改数据

使用 update_item()修改一个数据项目。使用 update_item()时必须指定要修改的项目的关键字 Key，同时指定要修改的数据项目，例如：

```
import boto3
import sys
#省略 readKeys()函数
try:
        keys=readKeys()
        dynamoDB = boto3.resource('dynamodb',
            aws_access_key_id=keys["keyID"],
            aws_secret_access_key=keys["secretKey"],
            region_name="cn-northwest-1")
        table = dynamoDB.Table("scenes")
        key={'sName': '双月湾风景区'}
        resp=table.get_item(Key=key)
        Item=resp["Item"]
        print("Before: ",Item["sLevel"])
        print("Before: ",Item["sScene"])
        table.update_item(
            Key=key,
            UpdateExpression='SET sLevel= :sLevelVal,sScene=:sSceneVal,sNote=:sNote',
            ExpressionAttributeValues={
                ':sLevelVal': '5A',
                ':sSceneVal': {'双月湾沙滩浴场':'免费','山顶观景台':'免费'},
                ':sNote':"风景独好"
            }
        )
        resp=table.get_item(Key=key)
        Item=resp["Item"]
        print("After: ",Item["sLevel"])
        print("After: ",Item["sScene"])
        print("After: ", Item["sNote"])
```

```
except Exception as err:
    print("errors: ", str(err))
```
注意，在修改时通过：
```
UpdateExpression='SET sLevel= :sLevelVal,sScene=:sSceneVal,sNote=
:sNote',
```
指定修改的数据是 sLevel、sScene 与 sNote。其中，":sLevelVal"":sSceneVal"":sNote"是形式参数，名称前面有一个"：",名称可以自由设置，具体的值通过下面的语句来设置：
```
ExpressionAttributeValues={
    ':sLevelVal': '5A',
    ':sSceneVal': {'双月湾沙滩浴场':'免费','山顶观景台':'免费'},
    ':sNote':"风景独好"
}
```
执行该程序，结果如下：
```
Before:  4A
Before:  ['双月湾沙滩浴场','海龟基地','山顶观景台']
After:   5A
After:   {'双月湾沙滩浴场':'免费','山顶观景台':'免费'}
After:   风景独好
```
由此可见，sLevel 与 sScene 的数据被修改了，而且 sScene 的数据类型由原来的列表变成了字典，同时增加了一个名称为 sNote 的数据项目。

### 6.4.4　删除数据

使用 delete_item() 删除一个数据项目。使用 delete_item() 时必须指定要删除的项目的关键字 Key，例如：
```
import boto3
import sys
#省略 readKeys()函数
try:
    keys=readKeys()
    dynamoDB = boto3.resource('dynamodb',
        aws_access_key_id=keys["keyID"],
        aws_secret_access_key=keys["secretKey"],
        region_name="cn-northwest-1")
    table = dynamoDB.Table("scenes")
    key={'sName': '双月湾风景区'}
    table.delete_item(Key=key)
    print("done!")
except Exception as err:
    print("errors: ", str(err))
```
执行该程序后显示"done!"，就表明数据被删除了。

### 6.4.5　扫描数据

使用 scan() 查找所有的数据项目，它返回的数据包含多个项目，例如：
```
import boto3
import sys
```

```python
#省略readKeys()函数
try:
    keys=readKeys()
    dynamoDB = boto3.resource('dynamodb',
        aws_access_key_id=keys["keyID"],
        aws_secret_access_key=keys["secretKey"],
        region_name="cn-northwest-1")
    table = dynamoDB.Table("scenes")
    table.put_item(
        Item={
            'sName': '双月湾风景区',
            'sLevel': '4A',
            'sSource': '国家级风景名胜区',
            'sType': ['海滨海岛', '生物景观'],
        }
    )
    table.put_item(
        Item={
            'sName': '深圳东部华侨城',
            'sLevel': '5A',
        }
    )
    resp=table.scan()
    Items=resp["Items"]
    for item in Items:
        print(item["sName"],item["sLevel"])
        if "sType" in item.keys():
            print(item["sType"])
except Exception as err:
    print("errors: ", str(err))
```

执行该程序，结果如下：

```
双月湾风景区 4A
['海滨海岛', '生物景观']
深圳东部华侨城 5A
```

其中，第二个景区"深圳东部华侨城"不包含 sType 的数据字段。

由此可感受到 DynamoDB 存储数据的灵活性，它不要求每个项目的数据结构是相同的，而一般的关系数据库中要求不同记录的数据结构是相同的。

### 6.4.6 删除数据库表

使用 delete()删除数据库表，例如：

```python
import boto3
import sys
#省略readKeys()函数
try:
    keys=readKeys()
    dynamoDB = boto3.resource('dynamodb',
        aws_access_key_id=keys["keyID"],
        aws_secret_access_key=keys["secretKey"],
```

```
        region_name="cn-northwest-1")
    table = dynamoDB.Table("scenes")
    table.delete()
    print("done!")
except Exception as err:
    print("errors: ", str(err))
```

执行完毕后,数据库表 scenes 就被删除了。

## 6.5 综合项目 爬取模拟景区网站数据

6-5-A
知识讲解

6-5-B
操作演练

### 任务目标

首先创建一个有两个景区的景区网站,编写爬虫程序爬取网站的数据,并把数据存储到 AWS 的 DynamoDB 数据库中。

### 6.5.1 创建模拟景区网站

使用 6.2 节中创建的模拟景区网站,爬取该网站的数据并将其存储到 AWS 的 DynamoDB 数据库中。

### 6.5.2 编写爬虫程序

编写爬虫程序 spider.py 来爬取网站数据,并将数据存储在 DynamoDB 数据库的 scenes 表中,程序如下:

```python
import urllib.request
import bs4
from bs4 import BeautifulSoup
import boto3
import time
import sys

class MySpider:

    def readKeys(self):
        keys={}
        try:
            f=open("accessKeys.csv","rt")
            rows=f.readlines()
            if len(rows)>=2:
                s=rows[1].strip("\n")
                s=s.split(",")
                keys={"keyID":s[0],"secretKey":s[1]}
            f.close()
        except Exception as err:
            print(err)
            sys.exit(0)
        return keys

    def __init__(self):
        try:
```

```python
            keys=self.readKeys()
            self.dynamoDB = boto3.resource('dynamodb',
                aws_access_key_id=keys["keyID"],
                aws_secret_access_key=keys["secretKey"],
                region_name="cn-northwest-1")
        except Exception as err:
            print(err)

    def openDB(self):
        try:
            try:
                self.table = self.dynamoDB.Table("scenes")
                self.table.delete()
                time.sleep(15)
            except:
                pass
            self.table=self.dynamoDB.create_table(
                TableName="scenes",
                KeySchema=[
                    {
                        'AttributeName': 'sName',
                        'KeyType': 'HASH'
                    }
                ],
                AttributeDefinitions=[
                    {
                        'AttributeName': 'sName',
                        'AttributeType': 'S'
                    }
                ],
                ProvisionedThroughput={
                    'ReadCapacityUnits': 5,
                    'WriteCapacityUnits': 5
                }
            )
            self.dynamoDB.meta.client.get_waiter('table_exists').wait(TableName="scenes")
        except Exception as err:
            print("errors: ",str(err))

    def insertDB(self,sName,sType,sSource,sLevel,sTime,sHotel,sScene,sRoute):
        #插入记录
        try:
            self.table.put_item(
                Item={
                    "sName":sName,
                    "sType":sType,
                    "sSource":sSource,
                    "sLevel":sLevel,
                    "sTime":sTime,
```

## 项目 ❻ 爬取景区网站数据

```python
                        "sHotel":sHotel,
                        "sScene":sScene,
                        "sRoute":sRoute,
                    }
                )
            except Exception as err:
                print("errors: ", str(err))

    def getFirstText(self,li):
        #获取第一个子节点
        for c in li.children:
            if isinstance(c,bs4.element.NavigableString):
                d=c.string.replace(" ","").replace("\n","")
                if d!="":
                    return d
        return ""

    def spider(self,url):
        #爬虫函数
        try:
            resp=urllib.request.urlopen(url)
            html=resp.read().decode()
            soup=BeautifulSoup(html,"lxml")
            lis=soup.find("ul").find_all("li",recursive=False)
            for li in lis:
                div=li.find("div",attrs={"class":"sightdetail"})
                sName=div.find("h3").text
                list=div.find("ul").find_all("li")
                sType=[]
                for sp in list[0].find_all("span"):
                    sType.append(sp.text)
                sSource=list[1].find("span").text
                sLevel=list[2].find("span").text
                sTime=list[3].find("span").text
                sHotel=[]
                for h in div.find("div",attrs={"class":"sighthotel"}).find_all("li"):
                    ht={}
                    ht["name"]=self.getFirstText(h)
                    ht["price"]=h.find("span").text
                    sHotel.append(ht)
                sScene=[]
                for sp in div.find("h6").find_all("div")[0].find_all("span"):
                    sScene.append(sp.text)
                sRoute=[]
                for sp in div.find("h6").find_all("div")[1].find_all("span"):
                    sRoute.append(sp.text)
                self.insertDB(sName,sType,sSource,sLevel,sTime,sHotel,sScene,sRoute)
        except Exception as err:
```

```
                    print(err)

    def show(self):
        #显示函数
        resp=self.table.scan()
        no=0
        for item in resp["Items"]:
            no=no+1
            print("No",no)
            print("名称: ",item["sName"]," 时间: ",item["sTime"],
" 级别: ",item["sLevel"]," 资质: ", item["sSource"])
            sType=item["sType"]
            print("类型: ")
            for t in sType:
                print("   ",t)
            print("酒店: ")
            sHotel=item["sHotel"]
            for h in sHotel:
                print("   ",h["name"],h["price"])
            sScene = item["sScene"]
            print("景区: ")
            for t in sScene:
                print("   ",t)
            sRoute = item["sRoute"]
            print("交通: ")
            for t in sRoute:
                print("   ",t)
            print()
        print("Total ",no)

#主程序
spider=MySpider()
spider.openDB()
spider.spider("http://127.0.0.1:5000")
spider.show()
```

### 6.5.3 执行爬虫程序

启动模拟景区网站的服务器，执行爬虫程序后得到如下结果。

```
No 1
名称： 双月湾风景区   时间： 四季皆宜   级别： 4A   资质： 国家级风景名胜区
类型：
    海滨海岛
    生物景观
酒店：
    金海湾喜来登度假酒店：¥700
    双月湾虹海湾假日酒店：¥400
景区：
```

```
        双月湾沙滩浴场
        海龟基地
        山顶观景台
交通：
        深汕高速-稔山出口-左转-大浦屯-右转-双月湾
        惠东-双月湾：约 20 分钟一班

No 2
名称：  深圳华侨城旅游度假区    时间：   四季皆宜    级别：   5A   资质：  国家级风景名胜区
类型：
        城市风景
        生态景观
酒店：
        雅枫宾馆（深圳华侨城店）：¥1700
        深圳华侨城洲际大酒店：¥1400
        深圳华侨城蓝汐精品酒店：¥2400
景区：
        欢乐海岸
        锦绣中华
交通：
        地处华侨城风景区，离华侨城地铁口仅 50 米；离市中心 15 千米，10 分钟车程
Total   2
```

程序执行完成后，打开 AWS 的 DynamoDB 数据库可以看到存储在 scenes 表中的数据，表中有两条记录，sHotel 的数据是字典数据，如图 6-5-1 所示。

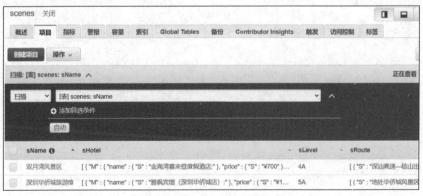

图 6-5-1  scenes 表中的数据

##  实战项目  爬取实际景区网站数据

### 🔍 任务目标

本项目的任务目标是从中国旅游信息网站中爬取 5A 景区的数据并将其存储到 AWS 的 DynamoDB 数据库中。

## 6.6.1 解析网站的 HTML 代码

使用 Chrome 浏览器访问网站,在网站中找到一个景区,将鼠标指针移至该景区,单击鼠标右键,在弹出的快捷菜单中选择"检查"命令,就可以看到图 6-6-1 所示的 HTML 代码。

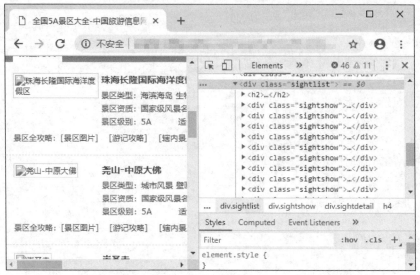

图 6-6-1　HTML 代码

从 HTML 代码可以看到景区的数据包含在<div class="sightlist">元素中,每个景区数据包含在一个<div class="sightshow">元素中。复制一个景区的 HTML 代码,整理后得到:

```
<html>
 <body>
  <div class="sightshow">
   <div class="sightimg">
    <a href="http://scenic.cthy.com/scenic-14115/">
     <img alt="珠海长隆国际海洋度假区" height="72" src="http://scenic.cthy.com/UploadPic/PhotoAlbum_Images/JingQu/635335033534313698.jpg" width="115"/>
    </a>
   </div>
   <div class="sightdetail">
    <h4>
     <a href="http://scenic.cthy.com/scenic-14115/" target="_blank">
      珠海长隆国际海洋度假区
     </a>
    </h4>
    <ul class="sightbase">
     <li>
      景区类型:
      <a class="ls" href="/scenicSearch/0-101-0-0-0-1.html" target="_blank">
       海滨海岛
```

```html
        </a>
        <a class="ls" href="/scenicSearch/0-105-0-0-0-1.html" target="_blank">
          生物景观
        </a>
      </li>
      <li>
        景区资质：
        <a class="ls" href="/scenicSearch/0-0-0-301-0-1.html" target="_blank">
          国家级风景名胜区
        </a>
      </li>
      <li>
        <span>
          景区级别：
          <a class="ls" href="/scenicSearch/0-0-201-0-0-1.html" target="_blank">
            5A
          </a>
        </span>
        适合季节：
        <a class="ls" href="/scenicSearch/0-0-0-0-405-1.html" target="_blank">
          四季皆宜
        </a>
      </li>
    </ul>
    <ul class="sighthotel">
      <li>
        <span>
          ￥230
        </span>
        <a href="http://hotel.cthy.com/hotelinfo-8498/" target="_blank" title="珠海海泉湾客栈">
          珠海海泉湾客栈
        </a>
      </li>
      <li>
        <span>
          ￥0
        </span>
        <a href="http://hotel.cthy.com/hotelinfo-68151/" target="_blank" title="珠海海泉湾海洋温泉中心">
          珠海海泉湾海洋温泉中心
        </a>
      </li>
      <li>
        <span>
          ￥700
```

```
            </span>
                <a href="http://hotel.cthy.com/hotelinfo-4564/" target="_blank" title="珠海海泉湾维景大酒店">
                    珠海海泉湾维景大酒店
                </a>
            </li>
          </ul>
        </div>
        <h6>
         景区全攻略：
            <a href="http://scenic.cthy.com/scenic-14115/photo.html" target="_blank">
                [景区图片]
            </a>
            <a href="http://scenic.cthy.com/scenic-14115/travelGuide.html" target="_blank">
                [游记攻略]
            </a>
            <a href="http://scenic.cthy.com/scenic-14115/attractions.html" target="_blank">
                [辖内景区]
            </a>
            <a href="http://scenic.cthy.com/scenic-14115/food.html" target="_blank">
                [美食特产]
            </a>
            <a href="http://scenic.cthy.com/scenic-14115/traffic.html" target="_blank">
                [交通概况]
            </a>
            <a href="http://scenic.cthy.com/scenic-14115/entertainment.html" target="_blank">
                [休闲娱乐]
            </a>
            <a href="http://scenic.cthy.com/scenic-14115/weather.html" target="_blank">
                [天气情况]
            </a>
        </h6>
    </div>
  </body>
</html>
```

## 6.6.2 爬取网站景区数据

爬取中国旅游信息网的网站，获取 HTML 代码，创建 BeautifulSoup 对象 soup：

```
url="http://scenic.cthy.com/scenicSearch/0-0-201-0-0-1.html"
resp=urllib.request.urlopen(url)
html=resp.read().decode()
soup=BeautifulSoup(html,"lxml")
```

我们使用 soup 对象爬取数据。

## 1. 爬取景区数据

景区数据都包含在<div class="sightlist">元素中,这个元素包含很多<div class="sightshow">,因此,使用:

```
divs=soup.find("div",attrs={"class":"sightlist"}).find_all("div",attrs={"class":"sightshow"})
```

循环 divs,得到每个<div class="sightshow">元素对象 div。

## 2. 爬取景区名称

景区名称包含在<div class="sightdetail">的<h4>中,因此使用:

```
dd=div.find("div",attrs={"class":"sightdetail"})
sName=dd.find("h4").find("a").text
```

爬取景区名称 sName。

## 3. 爬取景区类型

景区类型包含在<ul class="sightbase">下面的第一个<li>的各个<a>中,把各个<a>的文本连接在一起就得到景区类型 sType:

```
dd=div.find("div",attrs={"class":"sightdetail"})
lis=dd.find("ul",attrs={"class":"sightbase"}).find_all("li")
sType=[]
if len(lis)>0:
    for link in lis[0].find_all("a"):
        sType.append(link.text)
```

## 4. 爬取景区资质

景区资质包含在<ul class="sightbase">下面的第二个<li>的各个<a>中,把各个<a>的文本连接在一起就得到景区资质 sSource:

```
dd=div.find("div",attrs={"class":"sightdetail"})
lis=dd.find("ul",attrs={"class":"sightbase"}).find_all("li")
sSource=[]
if len(lis)>1:
    for link in lis[1].find_all("a"):
        sSource.append(link.text)
```

## 5. 爬取适合季节与景区级别

适合季节与景区级别包含在<ul class="sightbase">下面的第三个<li>的各个<span>中,因此适合季节 sTime 与景区级别 sLevel 可以这样获取:

```
dd=div.find("div",attrs={"class":"sightdetail"})
lis=dd.find("ul",attrs={"class":"sightbase"}).find_all("li")
if len(lis)>2:
    sLevel=lis[2].find("span").find("a").text
    sTime = lis[2].find("a",recursive=False).text
else:
    sLevel=""
    sTime=""
```

## 6. 爬取景区酒店

景区酒店的名称与价格等包含在<ul class="sighthotel">下面的<li>的各个<a>与<span>

中，因此景区酒店 sHotel 可以这样获取：

```
dd=div.find("div",attrs={"class":"sightdetail"})
lis = dd.find("ul", attrs={"class": "sighthotel"}).find_all("li")
sHotel=[]
for li in lis:
    h=[]
    h["name"]= li.find("a").text
    h["price"]= li.find("span").text
    sHotel.append(h)
```

### 6.6.3 爬取全部页面的数据

查看网页时发现：

第 1 页的地址是 http://scenic.cthy.com/scenicSearch/0-0-201-0-0-1.html；

第 2 页的地址是 http://scenic.cthy.com/scenicSearch/0-0-201-0-0-2.html；

……

第 n 页的地址是 http://scenic.cthy.com/scenicSearch/0-0-201-0-0-n.html。

并且在页面的底部可以看到共有 21 页，如图 6-6-2 所示。

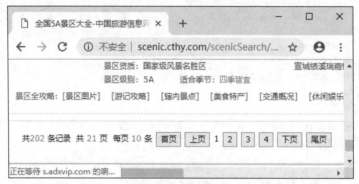

图 6-6-2　总页面数

该网站总页面数的 HTML 代码如下：

```
<ul id="PagerList">
    <li>
    共
    <span class="f14 point">
     202
    </span>
    条记录　共
    <span class="f14 point">
     21
    </span>
    页　每页
    <span class="f14 point">
     10
    </span>
    条
    </li>
```

```
...
</ul>
```

找到<ul id="PagerList">元素的第一个<li>元素下面的第二个<span>，就知道总页面数了。设计一个 getPageCount()函数来计算页面数：

```
def getPageCount(self):
    count=0
    try:
        resp=urllib.request.urlopen("http://scenic.cthy.com/
scenicSearch/0-0-201-0-0-1.html")
        html=resp.read().decode()
        soup=BeautifulSoup(html,"lxml")
        count=int(soup.find("ul",attrs={"id":"PagerList"}).find("li").
find_all("span")[1].text)
    except Exception as err:
        print(err)
    return count
```

### 6.6.4 设计存储数据库

程序使用 AWS 的 DynamoDB 数据库，数据存储在 scenes 表中。scenes 表的各个字段如表 6-6-1 所示。

表 6-6-1　scenes 表的各个字段

| 字段名称 | 类型 | 含义 |
| --- | --- | --- |
| sName | 字符串 | 景区名称（关键字） |
| sType | 字符串 | 景区类型 |
| sSource | 字符串 | 景区资质 |
| sLevel | 字符串 | 景区级别 |
| sTime | 字符串 | 适合季节 |
| sHotel | 字符串 | 景区酒店 |

### 6.6.5 编写爬虫程序

根据前面的分析，编写爬虫程序 spider.py，如下：

```
import urllib.request
from bs4 import BeautifulSoup
import boto3
import time
import sys

class MySpider:

    def readKeys(self):
        keys={}
        try:
            f=open("accessKeys.csv","rt")
            rows=f.readlines()
            if len(rows)>=2:
```

```python
                    s=rows[1].strip("\n")
                    s=s.split(",")
                    keys={"keyID":s[0],"secretKey":s[1]}
            f.close()
        except Exception as err:
            print(err)
            sys.exit(0)
        return keys

    def __init__(self):
        try:
            keys=self.readKeys()
            self.dynamoDB = boto3.resource('dynamodb',
                aws_access_key_id=keys["keyID"],
                aws_secret_access_key=keys["secretKey"],
                region_name="cn-northwest-1")
        except Exception as err:
            print(err)

    def initDB(self):
        try:
            try:
                self.table = self.dynamoDB.Table("scenes")
                self.table.delete()
                time.sleep(15)
            except:
                pass
            self.table=self.dynamoDB.create_table(
                TableName="scenes",
                KeySchema=[
                    {
                        'AttributeName': 'sName',
                        'KeyType': 'HASH'
                    }
                ],
                AttributeDefinitions=[
                    {
                        'AttributeName': 'sName',
                        'AttributeType': 'S'
                    }
                ],
                ProvisionedThroughput={
                    'ReadCapacityUnits': 5,
                    'WriteCapacityUnits': 5
                }
            )
            self.dynamoDB.meta.client.get_waiter('table_exists').wait(TableName="scenes")
        except Exception as err:
            print("errors: ",str(err))

    def openDB(self):
        try:
```

```python
                self.table = self.dynamoDB.Table("scenes")
        except Exception as err:
            print("errors: ",str(err))


    def insertDB(self,sName,sType,sSource,sLevel,sTime,sHotel):
        #插入记录
        try:
            self.table.put_item(
                Item={
                    "sName":sName,
                    "sType":sType,
                    "sSource":sSource,
                    "sLevel":sLevel,
                    "sTime":sTime,
                    "sHotel":sHotel
                }
            )
        except Exception as err:
            print("errors: ", str(err))


    def spider(self,url):
        #爬虫函数，用于爬取数据
        try:
            resp=urllib.request.urlopen(url)
            html=resp.read().decode()
            soup=BeautifulSoup(html,"lxml")
            divs=soup.find("div",attrs={"class":"sightlist"}).find_all("div",attrs={"class":"sightshow"})
            for div in divs:
                dd=div.find("div",attrs={"class":"sightdetail"})
                sName=dd.find("h4").find("a").text
                lis = dd.find("ul", attrs={"class": "sightbase"}).find_all("li")
                sType = []
                if len(lis) > 0:
                    for link in lis[0].find_all("a"):
                        sType.append(link.text)
                sSource=[]
                if len(lis)>1:
                    for link in lis[1].find_all("a"):
                        sSource.append(link.text)
                if len(lis)>2:
                    sLevel=lis[2].find("span").find("a").text
                    sTime = lis[2].find("a",recursive=False).text
                else:
                    sLevel=""
                    sTime=""
                lis = dd.find("ul", attrs={"class": "sighthotel"}).find_all("li")
                sHotel = []
```

```python
                    for li in lis:
                        h = {}
                        h["name"] = li.find("a").text
                        h["price"] = li.find("span").text
                        sHotel.append(h)
                    self.insertDB(sName,sType,sSource,sLevel,sTime,sHotel)
                    print(sName)
        except Exception as err:
            print(err)

    def showDB(self):
        #显示函数
        resp=self.table.scan()
        no=0
        for item in resp["Items"]:
            no=no+1
            print("No",no)
            print("名称:",item["sName"]," 时间:",item["sTime"]," 级别: ",item["sLevel"]," 资质: ", item["sSource"])
            sType=item["sType"]
            print("类型: ")
            for t in sType:
                print("   ",t)
            print("酒店: ")
            sHotel=item["sHotel"]
            for h in sHotel:
                print("   ",h["name"],h["price"])
            print()
        print("Total ",no)

    def getPageCount(self):
        #计算页面数
        count=0
        try:
            resp=urllib.request.urlopen("http://scenic.cthy.com/scenicSearch/0-0-201-0-0-1.html")
            html=resp.read().decode()
            soup=BeautifulSoup(html,"lxml")
            count=int(soup.find("ul",attrs={"id":"PagerList"}).find("li").find_all("span")[1].text)
        except Exception as err:
            print(err)
        return count

    def process(self):
        #爬取过程函数
        self.initDB()
        count=self.getPageCount()
        print("Total ",count," pages")
        #分页爬取
```

```
            for p in range(1,count+1):
                url= "http://scenic.cthy.com/scenicSearch/0-0-201-0-0-"+str(p)+".html"
                print("Page ",p," ",url)
                self.spider(url)
            self.showDB()

    def show(self):
        #显示数据
        self.openDB()
        self.showDB()

#主程序
spider=MySpider()
while True:
    print("1.爬取")
    print("2.显示")
    print("3.退出")
    s=input("选择(1,2,3):")
    if s=="1":
        spider.process()
    elif s=="2":
        spider.show()
    elif s=="3":
        break
```

这个程序定义了一个爬虫类 MySpider，其中有几个主要的函数。

（1）openDB()函数：用于创建数据库连接对象，建立数据库连接。

（2）initDB()函数：用于初始化 scenes 表，如果表已经存在，则调用 drop table 命令删除，然后调用 create table 命令创建表，以保证每次爬取数据时该表是空的。

（3）showDB()函数：用于显示数据库中的记录。

（4）spider()函数：这是程序的核心函数，用于爬取网页中景区的数据。

## 6.6.6 执行爬虫程序

执行爬虫程序，成功爬取了网站的 200 多个景区的数据。图 6-6-3 所示是在 AWS 的 DynamoDB 数据库中看到的 scenes 表的部分数据。

图 6-6-3　scenes 表的部分数据

## 项目总结

这个项目涉及一个景区网站，通过编写爬虫程序爬取了该网站的数据，并使用 AWS 的 DynamoDB 数据库实现了不规则数据的存储。

实际上，很多数据是不规则的，不同的数据行存在一些差异，使用传统的关系数据库存储这种类型的数据比较困难，而使用 NoSQL 数据库存储会简单很多。目前流行的 NoSQL 数据库有很多，AWS 的 DynamoDB 数据库就是常用的一种。

## 练习 6

1. AWS 的 DynamoDB 数据库与常用的其他数据库（如 MySQL、SQLite）相比有什么特点？

2. 如何理解 DynamoDB 数据库的数据项目，它与一般关系数据库的表格记录相比有什么不同？

3. 一组学生的选课情况如表 6-8-1 所示。

表 6-8-1 选课情况

| 姓名（Name） | 课程（Course） |
| --- | --- |
| A | Python、Java、MySQL |
| B | Python |
| C | 暂缺 |

完成下列任务。

（1）建立 DynamoDB 表格 courses 存储这组学生的选课情况，课程用列表表示。

（2）修改 B 同学的课程为"Java、JavaScript"，C 同学的课程为"MySQL"。

（3）查找所有选修"Java"课程的同学。

（4）如果已知 Python、Java、MySQL、JavaScript 课程的学分分别是 3、4、2、3，修改每位同学的课程数据结构为字典，如将 A 同学的课程数据结构修改为：

```
{"Python":3;"Java":4;"MySQL":2}
```

（5）统计每位同学的总学分。